MINOR FRUIT CROPS OF INDIA
OF INDIA
Tropical and Subtropical

The Author

Prof. Bibhas Chandra Mazumdar has been in the profession of teaching and supervising researches on fruit culture, growth and developmental physiology of crop plants and biochemistry of fruit crops in some Indian Universities since the year 1971 after his doctoral degree on biochemistry of growth, flowering and fruiting of mango cultivars. Major areas of his researches in teaching career include growth and developmental aspects of fruit crops, pectic metabolism, plant physiology, papaya latex and on less-exploited minor fruit crops grown in India. In the area on physiology and biochemistry of fruit growth, a number of his research contributions are considered to be of fundamental importance. He has so far to his credit 5 text books on fruit crops and horticulture and 121 research papers published in the journals of India, U.S.A. and U.K. as well as 17 scientific articles brought out in the magazines of India and the U.K. in the same areas.

MINOR FRUIT CROPS
OF INDIA
Tropical and Subtropical

Dr. B.C. MAZUMDAR

M.Sc. (Agri.); Ph.D.; F.I.A.B.S.; F.I.L.E.E.

Professor in Horticulture,
Institute of Agricultural Science, Calcutta University.
(*Formerly*, Reader in Horticulture,
Bidhan Ch. Krishi Vishvavidyalaya, Mohanpur, West Bengal;
Lecturer in Horticulture,
Allahabad Agricultural Institute, Allahabad, U.P.;
Fruit Development Officer, Government of West Bengal)

2012
Daya Publishing House®
A Division of
Astral International Pvt. Ltd.
New Delhi - 110 002

© 2004, Author

First Published, 2004

Reprinted, 2012

ISBN 978-93-5124-115-7 (International Edition)

Published by : **Daya Publishing House®**
A Division of
Astral International Pvt. Ltd.
– ISO 9001:2008 Certified Company –
4760-61/23, Ansari Road, Darya Ganj
New Delhi-110 002
Ph. 011-43549197, 23278134
E-mail: info@astralint.com
Website: www.astralint.com

Laser Typesetting : **Classic Computer Services**
Delhi - 110 035

Printed at : **Chawla Offset Printers**
Delhi - 110 052

PRINTED IN INDIA

Dedicated to the venerated memory of

Dr. BOSHISHWAR SEN
(1887 – 1971)

The founder of the *Vivekananda Research Laboratory* at Almora, whose pioneering researches in many areas of agricultural science are exemplary to the later scientists of India.

Preface

The study and culture of fruit crops in India had its inception in pre-historic times. The ancient sages and scientists in this land had pursued adequate researches to unearth the enormous wealth hidden in the fruits for use as food, medicine or other purposes and also how to render them more productive by meeting their specific requirements. Backed by this knowledge, they started setting up plantations of fruit crops near their dwelling places for close supervision of the trees, for provision of shelter to the domestic animals or for use as the sanctum to teach their disciples. Any fruit plant having any virtue had been introduced by them without making discrimination.

With rolling of ages, the necessity of food however, got dominance due to increasing population for which, fruits having greater suitability for use as food had been emphasized than those having other traits. The trend got impetus for land competition due to husbandry of other crops, reared animals and human settlement, eventuating the fruits having lesser importance as food to grow as wild. This is not the case only in India but also in other parts of the world. The fruits introduced in the plantations evidently, used to receive more and more care.

Later, the Moghuls on coming to this country were charmed with the fruit wealth and concentrated more on improvement of the fruits which have greater appeal and grown in the orchards. Similar line of action was taken by the Britishers as well, especially for the temperate fruits. On the other hand, the less cared fruits grown as wild tried to learn how to survive and produce in adverse situation and for this purpose, many of them gave birth to sub-species, varieties, hybrids or eco-types to suit the existing eco-environment. In course of time, man has however, discovered that many of these fruit crops, popularly recognized as the minor or the less-known fruit crops are equipped with a number of attributes apart from the demerits and hence, those need adequate exploration.

Out of a large number of such less-exploited, minor fruits that found to grow in different parts of India, only a few had been selected for discussion in the present text, which is an attempt to meet the requirements of the students and researchers in the areas on horticulture, applied botany, ayurved, forestry and related areas. Lastly, while gratefully acknowledging everybody who had assisted in this task, the author welcomes valued suggestions from the readers for improvement of the title in future.

Calcutta University **B.C. MAZUMDAR**

Contents

List of Plates

Minor Fruits and Their Importance

An exact definition of minor fruit crops is perhaps difficult. In a general sense, those fruits which though are consumable to the human beings but relatively less palatable than other fruits, which have lesser demand in the market, which are grown to a limited extent only and are not usually cropped in organized plantations with application of inputs are considered in grouping as minor fruit crops. Other terms that are used for these fruits are less-known fruits, less-appealing fruits, less-exploited fruits, stray fruits, wild fruits etc.

However, any sharp line of distinction between the major, *i.e.*, principal or more acceptable fruits and the less-known or minor fruits finds limitation if attempted to be done on a global basis. For example, mango is unhesitatingly ranked as the most important fruit in India while it is regarded as a minor fruit crop only in some western countries.

If area and production are considered to be the criteria to call a fruit crop as the principal or major as opposed to a less-known or minor, difference may be observed even in the same country as well. Edapho-climatic adaptability of fruit crops is beyond doubt, the pertinent factor that makes some to grow extensively over large area in a region unlike other parts in a country. Thus, the temperate fruits like apple, pear, peach, plum etc. are known to comprise as the major crops under the conditions of Kashmir and Himachal Pradesh while some varieties of those having less chilling requirement are although adaptable in the sub-tropics are grown to a small extent only in such situations. Similarly, tropical fruit crops like banana, jack-fruit, sapota, etc. are grown in Kerala and other parts of south India as the major fruit crops but these are grown in parts of north India sporadically.

Example may be cited as of the dry, sub-tropical situation of the Punjab, Haryana and adjoining region which has made it possible to grow Mediterranean grape over a large area yielding a high production unlike other parts. The moist, sub-tropical situation prevailing in the northern parts of Bihar and West Bengal has been on the other hand, the principal factor for extensive cropping of litchi, while the fruit is though grown in Dehradun and the Punjab but to a very small scale. The eastern part of India characterized by high rainfall and acidic soil offers congenial atmosphere for pine-apple that is grown as a major fruit crop there and which is though grown in the Pilibhit district and other parts of Uttaranchal is a minor fruit crop only. In the same way, mention may be made to the Nagpur region of Maharashtra where Mandarin orange ranks as the main fruit crop and the guava, which is considered as the major in the eastern U.P.

Apart from natural suitability, other conditions may also be responsible to occupy large area by a fruit crop in a given locality. The requirement of fruit crops in the industries may be pointed out such that existence of fruit preservation and processing factories in the nearness appears to be a principal cause for growing of some selected fruit crops there in many parts of India. Extensive cropping of papaya in many parts of south India is done by the growers especially to supply the latex tapped from the immature fruits to the nearby papain purification factories as the raw material. The taste of the people, in other words, their agreeability is also a factor that allocates some fruits to grow more in an area than the other parts of the country. For example, fruits having relatively high acerbity are usually preferred in south India, where even regular plantations of carambola, dillenia, tamarind etc. are met with at least in some parts but these are hardly set up elsewhere in the country. Mention may also be made to the fact that the taste of the people is likely to vary, *e.g.*, the grapefruit when first tried in the Punjab and other parts of north India had never been agreeable to the people because of somewhat bitterness in its juice but now has anchored well in that area of the country. Interestingly, the grapefruit is highly popular in the USA and the Europe mainly for the blend of bitterness in its juice due to the presence of naringenin and other bitter principles.

In many countries of the world, it is a noticeable feature that introduction of a new variety of a minor fruit crop having superior quality of fruit which is evolved by suitable breeding procedure has promoted it from the group of minor to the major or at least to the intermediate group. In China in particular, some varieties of excellent fruit quality have of late, been known to have developed in carambola, longan and other fruits and the area of those is increasing. In India also, this has become the reality for some fruits, thanks to the efforts made by the scientists. Perhaps Indian plum, *i.e.*, *ber* is a classical example in this country in this regard and a number of varieties or strains of this fruit tree having excellent organoleptic behaviour of fruits have been evolved during the past few decades. The fruits of these varieties have created adequate demand in the market and this has effectuated in setting up orchards of *ber* in many parts of the country. Accordingly, it is perhaps questionable whether to assign *ber* as a minor fruit crop in India now. Apart from *ber*, this has become possible for some other minor fruit crops as well under Indian condition.

From the foregoing discussion, it would become apparent that the term, minor fruit crop is a relative rather than a definite, besides that the list of such fruit crops in a country may also change with passage of time. However, in the area of horticultural science, the general trend is to consider the fruit crops grown in a particular country as a whole for making a classification of the principal or the major and the less known or the minor fruit crops. With that perspective, at least 150 species of human consumable fruits may be enlisted which could be placed in the group of minor fruit crops now in India considering the entire country as a whole. From that list, the present text has considered some selected tropical and sub-tropical species only for discussion.

Although organized orcharding is dispensed with for the minor fruit crops, many of those are merited with a number of attributes, important of which may be stated as follows.

Hardiness is by far, the most important quality for many of these fruit crops such that they are able to withstand adverse soil and climatic condition to a considerable extent. This is the reason why they multiply and grow spontaneously rather than coming to extinction. Yielding produce without receiving artificial agro-input is no doubt, a highly deserving character. Interestingly, some of these fruit crops are known to produce heavily every year and thus, the mechanism of utilizing the natural resource to the fullest extent is known to them. Attack by serious pests and diseases is relatively less for many of these fruit crops. Organoleptic behaviour of the fruits belonged to many of the minor fruit crops is beyond doubt, highly agreeable in *jamun*, *bael*, wax-apple, rose-apple and many other crops. A number of excellent products are also made from many of these fruits such as *morabba* from *aonla*, jam from *bael*, pickles etc. Many of these fruits have high nutritive value, *e.g.*, Barbades and West Indian cherries which rank as the richest source of vitamin C among all human consumable fruits and *aonla* ranks the next. Medicinal and aromatic properties of the fruits are known to a large number of such fruit crops. A number of species are attributed to have high timber value as well. These are also utilized as rootstocks or in breeding programme to evolve new varieties.

1

Aegle marmelos Correa

Common Names (English) : Wood-apple, Bael, Bengal quice.

Indian Names : Sreephal, Bilva, Bael, Bel, Bil, Vilvam, Bilpatre, Shool, Bela, Bilambu, Belli, Sriphalamu etc.

Botanical Family : Rutaceae.

Bael is the wonder tree of India. Among many uses of different parts of this tree, the following may particularly be mentioned:

(i) The ripe fruit is relished in fresh form. Although fruits of some varieties do not make good appeal for astringent taste due to the high content of phenolic compounds and also perhaps because of high mucilage as well as seeds, many other varieties are found to grow in different parts of this country, the fruits of which contain the above components in much lesser amount and have very thin rind such that, they are organoleptically much superior.

(ii) A number of preserved products are made from the fruits (Ray and Singh, 1978). The excellent *morabba,* especially of Varanasi may be mentioned first and besides that, jam is another highly palatable product (Mazumdar and Mukhopaddhya, 2003), large amount of which is now exported from India. Drying and grinding of the fruit is also common in some parts of India. Roy and Singh (1978) have stated the various ways of utilization of the fruits.

(iii) A number of medicinal properties of the fruits, leaves and roots of the tree are known since pre-historic times (Guha Bakshi et al., 1999). The therapeutic properties of the fruits in chronic amoebiasis, diarrhoea, gastro-intestinal ulcer, hyperglycaemia, haemorrhoids, costiveness, removal of intestinal worms and in other physical abnormalities have been amply mentioned (Jauhari and Singh, 1971; Roy and Singh, 1978; Alam et al., 1990). The crystalline substance, marmelosin extracted from the fruits has also been stated to have therapeutic properties. The roots are used in ethnomedicine and their ethanolic extracts had shown activity against many fungal and bacterial pathogens (Pitre and Srivastava, 1987). The leaves contain a number of volatile compounds having use in therapeutics and industry. Leaf extract has been reported to have anti-fungal properties against *Aspergillus* and *Penicillium* (Ramesh Tiwari et al., 1987; Alam et al., 1990). In having anti-aphrodisciac properties, the leaves of bael are taken by the Hindu *yogis* and saints in their worship.

(iv) Rind of the fruit has been reported to be a good source of natural dye (Khanna and Chandra, 1996).

(v) The tree serves as a protective barrier and is planted as an effective wind-break tree (Mazumdar, 2003).

(vi) Some varieties have graceful appearance of trees for which, those are planted in decorative gardens. Planting in the gardens is also done for pleasant smell of the flowers.

(vii) The role of leaves, fruit and wood of the tree need no emphasis in Hindu religion.

The tree of bael attains a height of about 8 metres and is deciduous. Leaves are trifoliate with oil glands. Branches are dimorphic and have straight, long and bold thorns. Flowers are bisexual, whitish green and have appealing smell. Petals are 5 in number but sometimes 4 and the ovary is ovoid. Kanchan Kumar Srivastava and Singh (2000) from Faizabad reported floral biology of bael. The fruit is characterized by a hard rind, whose thickness varies. Shape of the fruit ranges from oblong to ovate. Pulp is soft and yellow to orange in colour. Seeds are hard, numerous, compressed and surrounded by mucilage. The cotyledons are large.

Bael has been stated to be a native of India (Hayes, 1970; Pathak and Pathak, 1993) and specifically, north India (Ranjit Singh, 1969; Pareek and Sharma, 1993). In this country, the tree is abundantly grown in all sub-tropical and tropical parts and largely, in the north Indian plains. Apart from India, the tree is also grown much in other south Asian and south-east Asian countries and also in many middle-east countries. Of late, it is grown in some parts of the USA, Japan and other countries.

Basically, the tree is sub-tropical but many varieties are found to do exceedingly well in tropical situation of West Bengal, other parts of eastern India and also south India. The varieties or types which grow in sub-tropical climate usually do not make good performance when grown in tropical situation. In India, the tree grows upto an elevation of 1200 metres. Temperature as low as (-) 6.7°C is tolerated by the trees (Hayes, 1970). Soil is not considered to be an important factor and the tree grows in poor soil, either swampy or dry and tolerates alkaline soil (Hayes, 1970). Some varieties or perhaps types of bael are found to grow very successfully in the alkaline or saline-alkaline soil in the coastal Sunderbans area in West Bengal (Mazumdar, 1983).

In India, bael is largely propagated by seeds. This is the reason for so large variations that are observed in the tree and fruit characters in this country as the tree is highly heterozygous in its chromosomal make-up. The tree grows spontaneously here and there from seeds. Where planted, transplanting of seedlings is more common than direct sowing under Indian condition.

Mazumdar (1990) noted polyembryony in 33 per cent of the seeds procured from the south-central part of West Bengal and upto a maximum of 4 seedlings were found to be emerged from a seed. Increased longevity of the seeds by physical and chemical treatments (Hore and Sen, 1985, 1995), comparison of sowing media (Chattopadhyay and Mahanta, 1989) and effect of growth regulators and mechanical scarification on germination of seeds (Nayay and Sen, 1999) have also been studied under West Bengal condition.

With study done on stem cuttings, Ray and Chatterjee (1996) observed that growth regulators and etiolation treatments were significantly effective in inducing roots in ringed stem cuttings. Water shoots produced by invigoration enhanced rooting with etiolation treatment. Experiments done on air-layering by Mukherjee *et al.* (1986) provided results to show that invigoration, etiolation and application of IBA at 10,000 ppm gave maximum rooting response (90 per cent) with good tree survival (77.7 per cent).

High success (80 per cent) in budding done in June using budwood about a month old has been reported by Singh, L.B. as early as 1954. Patch budding, shield budding and chip budding are usually known to give good results in bael. Grafting is also successful and a study done by Maiti *et al.* (1999) showed that whip grafting was the most satisfactory method. Rootstock is seen to play a profound influence on scion and apart from *Aegle marmelos*, other species, *viz.*, *A. paniculata*, *A. chevlieri* etc. may also be used as rootstocks (Hayes, 1970).

A number of investigators have studied micro-propagation of bael. Varghese *et al.* (1993) in Gujarat used MS medium with 2, 4-D, NAA and kinetin. They observed that shoot buds that developed from nodal explants were most numerous when the medium was supplemented with kinetin and NAA. Islam *et al.* (1993) in Bangladesh had used leaf explants on MS medium supplemented with BA and IAA. Elongated shoots had been seen to root on half-strength MS with 0.1 mg IBA per litre. Hossain *et al.* (1994 b) from the same place had reported that callus-derived shoots produced roots and developed into plantlets when transferred to half-strength MS medium with 0.5 mg of IBA and 0.5 mg of NAA per litre. The study done by Islam *et al.* (1996) showed that the shoots which were regenerated rooted on half-strength MS medium with 25 mg of IBA per litre.

The changes in fatty acids, phospholipid and galactolipid contents during cellular and organ differentiation have been studied by Bhardwaj *et al.* (1995). Islam *et al.* (1996 a, b) had later studied on high frequency adventitious plant regeneration from radicle explants (1996 a) and adventitious shoot regeneration from root tips of intact seedlings (1996b). Arumugam and Rao (1996) had experimented on *in vitro* production of plantlets from cotyledonary node cultures using MS medium supplemented with BA and auxins. Morphogenetic responses of micro-shoots to benzyl amino purine had been studied by Hazarika *et al.* (1996) in Meghalaya. In Japan, plantlet regeneration from somatic embryos on MT medium containing 0.02 mg of NAA per litre had been induced by Ling Jing Tian and Iwamassa (1997). Hazarika *et al.* (1998) have later reported to have developed a protocol for rooting of bael micro-shoots *ex vitro.*

Organized orcharding of bael had not been a practice in India but lately, many growers are becoming interested and sizable plantations are found to have built up in parts of eastern U.P., Sunderbans area of West Bengal and other parts of the country. If planted, a spacing of 10-12 metres appears to be satisfactory. Singh and Roy (1984) have provided cultivation method of the tree.

De *et al.* (1996) had assessed the nitrogen, phosphorus and potassium status of the leaves of bael grown in the southern part of West Bengal. Analysis of the leaves sampled at monthly interval from the month of December to May had revealed a consistent increase of the elements except in case of potassium which showed decrease from April to May. Development of the fruits during the summer season had a concomitance with lesser increase of the components in the laminar tissues than in the winter season.

Fruiting in bael starts from the fifth year but commercial harvest is hardy possible before the seventh year. Grafted trees usually produce early. Harvesting should be done by plucking the fruits individually and preferably, before they attain full ripening. The fruits of bael are climacteric and although they ripen best in the tree, harvesting should be done slightly earlier to avoid natural dropping which may cause their cracking. The yield of fruits varies greatly and may be 200-500 fruits per tree.

A number of varieties have though been mentioned by some authors but many of those could better be regarded as types or eco-types rather than varieties. As a matter of fact, a large number of variations are observed in tree and fruit characters of bael because of seed propagation. However, Singh described 6 varieties from U.P. and among those, Mirzapuri variety was considered best. Jauhari et *at.* (1969) described physico-chemical characters of 7 genotypes collected at Fruit Research Station at Basti, U.P. According to them, Etawah kagzi, Sewan large, Mirzapuri and Deoria were excellent in quality. Roy and Singh (1978) studied 24 genotypes from Agra, Kolkata, Delhi and Varanasi and put forward that larger fruits had better processing quality. With survey of varieties grown in U.P. and Bihar, 13 types had been collected at Narendra Deva University of

Agriculture and Technology, Faizabad. Among those, the Narendra Bael (NB) 1, 5 and 6 were of superior quality.

The types of bael grown in the southern and central parts of West Bengal have been studied by Mazumdar and co-workers. In earlier study, 5 types of fruits differing in size and shape were analyzed (Mazumdar, 1975). Differences in weight and size of fruits, shell thickness, seed number and size, sugars and titratable acidity (equivalence of citric acid) were observed and the type having spherical flattened fruits was considered superior. Subsequently, 6 types of bael grown in the South 24-Paraganas (Sunderbans) district were analyzed by Kumri *et al.* (1997), in respect of shape, colour, weight, specific gravity, length and diameter of the fruits, thickness and weight of rind, seed number and weight as well as total, reducing and non-reducing sugars, titratable acidity, vitamin C and the pectin as calcium pectate in the pulp. The fruits differed widely in shape and great variation in regard some components was found to be apparent in some types.

In later study (Mukhopadhyay *et al.*, 2002), 3 types of bael procured from several plantations located in the Sunderbans (South 24-Paraganas district) area of West Bengal were assayed for shape, weight, specific gravity, length and diameter of fruits, rind thickness, pulp colour, seed weight and the content of dry matter, total, reducing and non-reducing sugars, titratable acidity, pectin (Ca-pectate) and vitamin C content in the pulp. Fruits grouped as of type "A", characterized by oblong shape was assessed to be superior for many characters. Pectin content in the pulp of fruits of some types grown in other parts of West Bengal has also been studied (Pratima Roy and Mazumdar, 1989; Ali and Mazumdar, 1991). Analysis of some types of bael grown in West Bengal have been done by Maity *et al.* (1999 b) as well.

It is of interest to add that a type is known to be grown in the western part of West Bengal, the rind of which is so thin as breaks easily by giving pressure with thumb. The type has been named as Kagzi, exaggerating that the rind is like paper is seen to have less mucilage in the pulp.

Jaiswal *et al.* (1999) studied the clonal variations in growth and yield of 8 years old budded trees of bael at Pantnagar, U.P. The results showed great variation in growth and yield attributes on various clones. Pant Bael-10 and 1 were noted to produce more fruits and Pant Bael-5 produced highest fruit weight per tree followed by Pant Bael-12. In the same place, Misra *et al.* (2000) evaluated early performance of 8 selected genotypes and tabulated data on tree and fruit characters.

Analytical reports of bael fruits as presented by Gopalan *et al.* (1993) is as follows. Moisture-61.5 per cent, protein-1.8 per cent, fat-0.3 per cent, minerals-1.7 per cent, fibre-2.9 per cent, carbohydrate-31.8 per cent, calcium-0.085 per cent, phosphorus-0.05 per cent and iron-0.6 mg per 100 g. Among vitamins, carotene-55 mcg, thiamine-0.13 mg, riboflavin-0.03 mg, niacin-1.1 mg and vitamin C-8 mg per 100 g of pulp have been stated by them. A report by Barthakur and Arnold (1989) pointed out that the bael fruits that were sent from India to Canada had high TSS and out of 17 amino acids detected, aspartic acid constituted 32 per cent. The said authors added that the iron content in the pulp was also high. Although Gopalan *et al.* (1993) had stated very low amount of iron, the materials used by these authors were thus, superior in respect of iron content.

The study pertaining to fruit development of bael carried out by Singh, R.D. (1986) at Saharanpur, Uttaranchal provided evidence that maximum development of the materials had occurred upto the beginning of October after which it slowed down until December and remained constant upto the first fortnight of April. Pande *et al.* (1986) in Kanpur observed that the sugars content had continued to be increased in the pulp. The acidity level was noted to be very low and it did not change much during development but the vitamin C content was found to have increased with maturity. Besides, sharp increase in pectin, tannin and marmelosin contents in the pulp was also observed until January following which, the components gradually declined. Ghosh *et al.* (1985) had assayed gibberellins in the developing fruits of bael.

A number of insect pests and diseases are found to attack bael. Fruit canker due to the bacterium, *Xanthomonas bilvae* is sometimes serious which produces water-soaked spots on the leaves. Other parts of the tree and fruits are also affected (Patel *et al.*, 1953). Ranjeet Singh *et al.* (1996) observed that aqueous extract of *neem* seed kernel (0.5 per cent) was effective in protection against the pest *Papilio demoleus* in the nursery seedlings by making spray at interval of 8 days. Damage to the rind should be avoided as this may cause pathogenic injury to the pulp.

Post-harvest storage is an important aspect in bael and pathogenic infection becomes very serious in some varieties in ambient storage. Dipping the fruits in lime water for a few minutes just after harvest is a practice done in many parts of India to avoid infection. Applying lime over the cut surface of the fruit-stalk is also effective to ward off infection in storage.

2

Anacardium occidentale Linn. (Cashew-apple)

Common Names (English) : Cashew, Kaju.

Indian Names   : Kajutaka, Kaju, Kaju badam, Hijli badam, Mindiri, Jidi-mamidi, Garo-biju, Lanka badam, Kashumavu etc.

Botanical Family   : Anacardiaceae.

In cashew, two parts have edible value to man. These are the kernels, *i.e.*, the semi-lunar and large-sized cotyledons which are present inside the shell and the cashew-apple, *i.e.*, the swollen, juicy peduncle attached to the proximal end of the nut. (In Java, young leaves are also cooked with rice to add flavour but this is of lesser importance).

However, the kernels present in the nuts are no doubt, the most important materials. These are highly relished on roasting with salt and spices. Besides, these are used as ingredient in chocolates, confectioneries and sweatmeats. The kernel has nutritive value especially in respect of good quality protein, carbohydrate, fat, some minerals and B-vitamins. In the international trade of tree nuts, it ranks third in contributing 20 per cent of all nuts and above all, India earns a lot of foreign money by exporting the kernels to other countries.

Despite the fact that the cashew kernel has a number of merits, the cashew-apple cannot be connived at. Botanically, cashew-apple is however, not a fruit because it does not develop from the ovary. The organ which is a pseudocarp, is in fact, the swollen form of the peduncle and the thalamus. But it develops a soft and juicy texture, appealing sweat-sour taste and flavour, attractive colour and shape and hence, from the horticultural point of view, it is grouped within fruits.

Cashew-apple has no doubt, much importance to man. The fruits are taken afresh but more importantly, a fermented drink, known as *feni* is prepared from its juice, especially in the Goa and Konkan region in India and also in some other countries, which has high popularity. Besides, good quality jam and squash are prepared from this material. From the nutritional point of view, cashew-apple is very rich in vitamin C. Medicinal value including anti-vomitting and anti-catarrhal effect of it is also known (Madhavan Rao, 1969).

However, the present text has confined discussion on cashew-apple only which is considered in the tropical parts of India as a minor fruit and any deliberation on cashew nut has been omitted.

It may be mentioned in this context that recorded researches on cashew-apple have not been as extensive as in case of cashew nut. Nevertheless, the pertinent works done on this minor fruit in the recent years have been brought forward for discussion.

As regards impact of culture on the cashew-apple, the works done by Kumar and Sreedharan (1987), Egbekun and Otiri (1999) and Akinwale and Aladesua (1999) may be brought to notice. Kumar and Sreedharan (1987) made a correlation study between leaf nutrients and fruit (apple) quality characters. They had applied N (150-450 g), P_2O_5 (50-150 g) and K_2O (50-160 g) per tree per year. A negative correlation was observed by them between laminar N, P, K content and fruit sugar and juice reaction, while a positive correlation with juice and vitamin C content.

The effect of different levels of N, P, K and plant growth regulators (2, 4-D, NAA, ethrel) on yield and quality of apple was tested by Egbekun and Otiri (1999) and it was noted that the most effective treatments were N (500 g) + P (250 g) + K (250 g) per tree and spraying ethrel of 50 ppm solution. Akinwale and Aladesua (1999) studied on the effect of plant spacing but the 3 spacings maintained by them had not been seen to have any significant difference in regard performance of the cashew-apples.

The yield of cashew-apple has been stated by Madhava Rao (1967) as 35 kg per tree. But it varies on many factors and largely on varieties or types. The yield potential of pink and yellow types of apples of a cashew cultivar has been evaluated by Owaiye (1996) in Nigeria and the pooled yield of 4 years obtained by them showed that yellow-coloured apple was higher yielder. Physico-chemical study on the red and yellow coloured apples had also been done by Akinwale and Aladesua (1999) in Nigeria with local and Brazilian types. They indicated that steaming of apples though improved palatability but reduced vitamin C content in the materials. Varietal evaluation of cashew-apple had later been studied by Moura *et al.* (2001) in respect of physical characters intended for fresh fruit market.

A number of reports are available in regard physico-chemical characteristics of cashew apple. Gopalan *et al.* (1993) stated that the cashew fruits contained 86.3 per cent moisture, 0.2 per cent protein, 0.1 per cent fat, 0.2 per cent minerals, 0.9 per cent fibre, 12.3 per cent carbohydrate, 0.01 per cent calcium, 0.01 per cent phosphorus and 0.002

per cent iron. Besides, 23 mcg of carotene, 0.02 mg of thiamine, 0.05 mg of riboflavin, 0.4 mg of niacin and 180 mg of vitamin C per 100 g of pulp has also been stated by them.

Maikup *et al.* (1997) had physico-chemically analyzed apples of three colours, *i.e.,* pure red, pure yellow and admixture of red and yellow which are generally found to be grown in the coastal area in the Medinipur district of West Bengal. The red coloured apples were estimated to have maximum values in respect of weight (63.6 g), reducing and non-reducing sugars (9.2 and 0.9 per cent), titratable acidity (0.38 per cent as equivalence of citric acid), vitamin C (258.3 mg per 100 g) and calcium pectate (2.11 per cent).

Evaluation of cashew apples of 9 early dwarf clones had been done by Moura *et al.* (2000) and they pointed out that the tannin content was in general, lesser than what had actually been reported in the literature. The study done by Olveira *et al.* (2002) in Brazil showed alanine, serine and other free- amino acids in the pulp. The volatile compounds present in cashew apple have been studied by some workers. Pino (1997) in Cuba made a review of such compounds and Bicalco and Rezendo (2001) reported esters, terpenes, hydrocarbons, alcohols, ketones, lactones and some other compounds.

In cashew, the growth of nut is noted to be faster than the apple at the early stage. But when the nut growth almost ceases after a month or so, the apple takes up quick growth and even outgrows the nuts. A number of workers have studied changing pattern of different physico-chemical components in the apples with progress of their development and ripening. Under West Bengal condition, Roy and Mazumdar (1989) had sprayed 0, 1, 3 or 5 per cent solutions of zinc sulphate to the trees when the nuts that had been suitably marked had attained 5-7 days of maturity. The marked nuts including the apples were analyzed after 20, 30 and 40 days and it was observed that the sugars, acidity and vitamin C content in them had enhanced by zinc sulphate solutions and directly with concentrations. Similar study was done by Bera and Mazumdar (1992) in the same experimental condition by spraying the trees with 0 or 200 ppm solutions of IAA, IBA or NAA. All the auxins were observed to have raised the level of vitamin C and acidity but not sugars. The responsiveness was however, maximum with IBA and even spraying with only water had produced some good results when the samples were analyzed after 20, 30 and 40 days of spraying.

In an attempt to study whether application of penicillin could augment the protein content of the cashew kernels, probably mediating through penicillamine, Bhattacharyya and Mazumdar (1993) had sprayed 0, 0.1, 0.5 or 0.7 per cent aqueous solutions of benzyl penicillin to the trees following similar experimental procedure as stated above. Along with kernels, the apples were also analyzed at different maturity condition. The general trend of results showed that weight and specific gravity of the apples had increased over control by spraying with 0.1 per cent penicillin solution, while higher concentrations were ineffective or had even lowered. All the concentrations had raised total sugars but maximum by 0.1 per cent solution. Vitamin C had also increased by this concentration while the higher concentration had lowered it. It seems logical to opine that effectiveness of penicillin may be there only at very low concentrations. This is also applicable as regards heightening of protein, carbohydrate and lipid matter of the kernels which is observable from their findings.

Naidu *et al.* (1998) in Andhra Pradesh studied dry matter accumulation in the apples during development and it was observed that a slow increase in dry weight had occurred until the 4th week and this was followed by a marked increase. Analyses in respect of many components in the apples during their development have also been reported by other workers, including Egbekun and Otiri (1999) and Figueiredo *et al.* (2001, 2002).

Bhattacharyya *et al.* (1989) reported the problem of cracking of the nuts in a large number of cashew trees grown in the southern coastal area of West Bengal. Such type of cracking which was noticed especially in later part of development of the nuts had resulted in lesser growth of apples as well which became under-sized. Filgueuras *et al.* (1999) studied on perishability of the apples and put forward suitable handling techniques for extending storage life of the apples.

Lastly, it may be concluded that adequate research on improvement of cashew apple especially to render those more suitable in making *feni* and other products has no doubt, a necessity in India.

3

Annona reticulata Linn.

Common Names (English) : Bullock's heart, Bull's heart.
Indian Names   : Krishnabija, Ramphal, Nona,
                          Nona   ata,   Ramasita,
                          Ramapandu, Anoda etc.
Botanical Family   : Annonaceae.

Bullock's heart is grown as wild in India but in some parts of West Bengal, it is planted for utilization of waste land, particularly in heavy soil and fruits have also demand there. Although the fruit is less sweet than custard apple and the pulp is not delicately flavoured, the carpels are fused giving a buttery consistency and seeds are lesser than custard apple. Sometimes the pulp of this fruit is mixed with that of other fruits for enhancing taste and flavour. The fruits of *A. reticulata* do not compete with custard apple as the former are available in summer and the latter in autumn season. The plants are also used as rootstocks for custard apple.

The tree of bullock's heart is deciduous, reaches upto 7-8 metres and more in rich soil. Leaves are oblong lanceolate with acute apices. The fruit is heart-shaped and the rind is marked with hexagonal areoles (Venkataratnam, 1965; Hayes, 1967).

Bullock's heart is known to be a native of tropical America and in India, it is grown as wild in all tropical parts. West Bengal is important for growing this fruit mostly as wild and apart from that, it is also abundantly found in Assam and other eastern States as well as in most parts of south India.

The requirements of climate and soil for bullock's heart are similar to custard apple but it is somewhat less resistant to cold or heat. It also comes up better in heavier soil (Venkataratnam, 1965). The trees are found to thrive upto 1300 metres in tropical situation but fruiting is not good in high attitude.

Propagation is done by seeds but budding on custard apple is also successful. Cartagena Valenzuela and Barreto Osoria (1998) reported improvement in seed germination and growth of seedlings of *A. reticulata* by application of gibberellic acid. The performance of *A. reticulata* plants grafted on eight rootstocks by adopting side grafting and oblique wedge methods have been evaluated by Marcelino Ponce (1986) in Mexico and it has been reported that percentage take was maximum (100 per cent) as well as plant growth was best in *A. reticulata* when side-grafted on *A. muricata, A. reticulata* and *A. lutescens.* Success in air-layering of *A. reticulata* by ringing of shoots and use of suitable concentrations and type of auxins have also been reported by Chovatia and Singh (2000 b) in Gujarat.

Gholap *et al.* (2000 a) in Akola, Maharashtra compared three budding methods in *A. reticulata* and reported that success was highest in softwood grafting (91.5 per cent), followed by patch budding (89 per cent) and shield budding (80.5 per cent).

While studying leaf sampling technique for nutritional diagnosis, Dhandar and Bhargava (1993) stated that sampling tissue from the 5th youngest leaf gave the best index of nutritional status in *A. reticulata*. The tree grows as wild in India without any agro-input but it is seen to yield tremendously by application of fertilizers and response of potash is very high to improve quality of fruits (Mazumdar, B.C., unpublished). Zang and Xu (2000) with pot-culture experiments done in China observed that nitrogen fertilizer was most beneficial for *A. reticulata* and this was followed by potassium and calcium, although phosphorus and magnesium had much less effect.

A. reticulata trees flower during August to October but fruits take long time, *i.e.*, eight months to mature and are available from March to May (Venkataratnam, 1965). In some tropical areas, fruits are available throughout the year. The yield varies from 25-50 fruits per tree and each fruit may weigh 250-350 g.

Any standard variety of A. *reticulata* is not known. According to Sham Singh *et al.* (1967) two types of fruits are observed of which one is the squamosa form with heart shaped appearance and the other one is reticulate form having distinct finger prints on surface. It has been stated that reticulate form bears more than the other.

According to Gopalan *et al.* (1993), the fruits contain 76.8 per cent moisture, 1.4 per cent protein, 0.2 per cent fat, 0.7 per cent minerals, 5.2 per cent fibre, 15.7 per cent carbohydrate, 0.010 per cent calcium, 0.010 per cent phosphorus and 0.006 per cent iron. A number of authors have also reported many volatile and other compounds from the fruits (Islam, 1986; Oguntimein, 1987; Wong and Khoo, 1993; Pino *et al.*, 1998; Pino, 2000). Ohsawa *et al.* (1990) from Japan reported insecticidal properties from ether extract made from the seeds of *A. reticulata*.

The changing responses of the constituents of *A. reticulata* fruits during development had been studied by Bhattacharyya *et al.* (1992) with fruits borne by trees in the

South 24-Paraganas district of West Bengal. The total soluble solids, titratable acidity, total, reducing and non-reducing sugars and the vitamin C content were analyzed, sampling fruits at three stages of development, *i.e.*, later part of April, early part of May and at harvest-maturity condition which was early part of June. All the constituents were observed to have increase with development except acidity which decreased. The increases were very high from early April to early May. Other informations have also been provided.

4

Artocarpus lakoocha Buch.
(*Artocarpus lakoocha* Roxb.)

Common Names (English)	: Monkey jack, Monkey jack-fruit.
Indian Names	: Dahu, Barhal, Deophal, Lakuch, Chama, Dewa, Jeuta, Deheo, Lahu, Lovi, Irappala, Lakuchamu, Kaunagona, Wotomba, Kammaregu, Votehuli, Daua etc.
Botanical Family	: Moraceae.

The fruits of monkey jack are small, *i.e.*, 7-10 cm in diameter, irregular in shape and as a fresh fruit, it is usually not very popular. But good quality pickle is prepared from the fruits along with seeds and a stiff jelly can also be prepared. The tree is handsome (Hayes, 1970) and sometimes planted along avenues. It is also selected for planting in informal style of gardens in many tropical countries.

The nativity of the tree is India or India to Malacca (Pareek and Sharma, 1993; Ranjit Singh, 1969) and according to Hayes (1970), in Bengal. Tropical climate is favourable for the tree but it also grows in the sub-tropics.

In India, the tree grows as wild, abundantly in the northern and central parts of West Bengal, Assam and other plain lands in the eastern States. It is also grown in the coastal areas of Kerala, Maharashtra, Tamil Nadu and other parts of south India. However, the tree thrives well in the sub-Himalayan areas where humidity is high.

Where planted, seedlings are only used, but air-layering is also successful when done in rainy-season with application of 0.1-0.2 per cent IBA in lanolin paste.

The fruit is climateric and although ripens best in the tree, is harvested slightly earlier to avoid softening.

No standard variety of monkey jack is known but a survey done in various parts of West Bengal revealed a number of types in respect of fruit shape, colour, shape and size of leaves and other characters (Mazumdar, B.C., unpublished).

According to Gopalan *et al.* (1993), the fruits contain 89.6 per cent moisture, 1.6 per cent protein, 1.2 per cent fat, 1.1 per cent mineral matter, 2.8 per cent fibre, 13.9 per cent carbohydrate, 0.067 per cent calcium and 0.025 per cent phosphorus.

5

Averrhoa carambola Linn.

Common Names (English) : Carambola, Star-fruit etc.

Indian Names   : Kamaranga, Kamrakh, Kamranga, Karmal, Karamara, Kordoitenga, Tamarta, Irimpanpuli etc.

Botanical Family   : Oxalidaceae.

The fruits of carambola are not taken as raw because of high acerbity but fruits of some types which are less sour and to some extent sweet are taken afresh by making slices and mixing with sugar and salt. However, the fruits have a demand to produce a number of preserved products notably, pickles, jam, jelly, preserve, drink etc. In south India, carambola fruits have high popularity and there it is used as a substitute of tamarind in cooking. It is much in use in West Bengal also to make *chutney*. In China, fruits of sweet taste are also known to be grown. A fragrance which is like that of quince is obtained in fruits of some types. It is said that the sour pulp of the fruits is used in removing stains from linen and in shining brass (Hayes, 1970). The carambola tree is planted along avenues and gardens as well for attractive and star-shaped appearance of the fruits and shape of the trees.

The trees attain a height of 8 to 10 metres and have symmetrical habit of growth. The leaves are odd-pinnate with leaflets numbering 5 to 9 and these increase in size towards the tip. They are found to be sensitive to touch and light and get folded when touched or in low light and darkness. The fruits are green to light yellow in colour and in some types, rich yellow colour is attained when fully ripe. The fruit has 3 to 5 deep longitudinal, prominent ribs such that it is star-shaped on cross-section. The skin is thin, smooth, translucent and the shape of the fruit is oval or more commonly, ovoid.

The origin of *A. carambola* is not clearly known. However, Ranjit Singh (1969) has mentioned the origin in India-China. Some authors are of opinion that the species is of Malayan origin. The trees are grown in all tropical and sub-tropical situations in India where there is very little frost and a warm, moist climate is congenial for better production (Hayes, 1970). The tree is largely grown on lower hill-slopes of south India upto an altitude of 1000 metres (Ranjit Singh, 1969) and in the west-coast.

Seed propagation is most common for *A. carambola* but vegetative propagation methods are also done in many countries. With study done on storage and germination of seeds, Sekiya and Cunha (1999) stated that the best emergence percentage and seedling vigour occurred when seeds were dried in the sun with or without the aril.

The use of auxin in suitable concentrations to induce rooting of cuttings has been stated by Bora and Das (1998). Grafting and especially, inarching, shield budding and Forkert budding are also done in many countries over the seedlings of the same species (Hayes, 1970). Marler and Mickelbart (1992) stated that GA could be suitably used to shorten nursery time for producing graftable rootstocks.

A number of reports are available on micro-propagation of *A. carambola*. Kantharaj *et al.* (1992) described a technique for micro-propagation using MS medium supplemented with BA and NAA. Methods for regenerating plants *in vitro* from callus cultures have also been put forward by Amin and Razzaque (1993) from Bangladesh. Li Ji Hong *et al.* (1999) with works done in China had mentioned that the best medium for shoot tip culture was MS supplemented with BA and NAA.

From the study done on accumulation of nitrogen, phosphorus and potash in the leaves of *A. carambola* sampled at monthly interval from December to May, it was observed by De *et al.* (1993) under West Bengal condition that all the nutrients had decreased with advancement of summer following increase in the winter months. Improvement in the quality, colour development and storage of the fruits by foliar application of the solutions of potassium sulphate, zinc sulphate and magnesium sulphate has also been observed (Mazumdar, B.C. *et at.,* unpublished).

No standard variety of *A. carambola* is available but a large number of variations in regard colour, shape, sweetness and sourness of the fruits are met with. Some of the Chinese strains claim to be very sweet and in Columbia, an outstanding type is Icambola (Pareek and Sharma, 1993). Ping Sheng (1999) had mentioned a high quality variety in China which is a sport of a sweet variety.

The fruits of *A. carambola* are largely harvested in the winter and autumn seasons and 50 to 100 kg of fruits are produced by a tree. The yellowish fruits usually have a greater demand in the market. Bhattacharyya *et al.* (1988) had reported degreening and yellowing of the harvested fruits in West Bengal by steeping with suitable concentrations of salt water. On steeping in 1 to 24 per cent solutions, the weight, specific gravity, texture, TSS, acidity and the vitamin C content as well as the extent of yellowing and degreening in the fruits have been determined by them. Ether induced degreening and yellowing as well as changes in the above physico-chemical components of the fruits had also been a study done by Bera and Mazumdar (1991).

The components of the *A. carambola* fruits had been determined by a number of workers. Heredia *et al.* (1998) observed that sucrose was the main carbohydrate in the fruits. The fruits have been reported to contain upto 4 per cent pectin by Pratima Roy and Mazumdar (1989). Mandal and Mazumdar (1995) reported average weight, specific gravity and length of fruits as 71 g, 0.097 and 8.1 cm respectively while in the pulp, the total, reducing sugars, non-reducing sugars and the titratable acidity (citric acid equivalence) had been respectively estimated as 2.84 per cent, 1.98 per cent, 0.82 per cent and 1.28 per cent. Pino (1997) reported the volatile compounds existent in the fruits. The effect of foliar spraying the trees with 1 per cent solutions of KCl or $ZnSO_4$, or 50 ppm solutions of IBA or GA_3 on weight, specific gravity, size as well as dry matter, total, reducing and non-reducing sugars, pectic compounds, titratable acidity and vitamin C content of the fruits had been reported by Sinha Roy (1996), Mandal (1997) and Mandal and Mazumdar (2000).

The physico-chemical components of fruits harvested at different maturity conditions in *A. carambola* have been determined by Siti Halijah Ali and Jaafar (1992). According to them, the TSS, *p*H, sugars and ascorbic acid increased with maturity but acidity level decreased. Mandal (1997) and Mandal and Mazumdar (1998) observed that the total duration of fruit development of *A. carambola* had been around 56 days in West Bengal. They had determined the physico-chemical components as mentioned above at four stages of development of the fruits at an interval of 14 days. Except acidity, all the constituents were noted to have increased and markedly from 42 days after fruit set till attainment of harvest-maturity, *i.e.*, 56 days after fruit set.

The acidity level had increased highly upto 42 days and thereafter decreased. The changing pattern of pectic fractions during development had also been studied by Sinha Roy (1996).

Post-harvest storage of *A. carambola* fruits had also shaken attention of many workers (Sankat and Balkisson, 1992; O' Hare, 1993; Miller and McDonald, 1997) who suggested measure to enhance shelf-life and improvement in packaging of the fruits.

6

Borassus flabellifer Linn.

Common Names (English) : Palmyra palm.
Indian Names   : Taruraja, Tal, Tad, Pana,
                               Karimpana, Talam, Talo,
                               Tanlo, Panai-marom, Chettu,
                               Tooti, Tamar, Taalimara,
                               Thadi, Tala etc.
Botanical Family   : Palmae.

Palmyra is the most widely growing palm in tropical parts of India and almost every part of this tree has utilitarian value to the human beings. The sweet sap or toddy is by far, the most important by-product having large use in this country. The spadices of the palm are tapped to secure this sap, *i.e.*, the syrupy liquid. Both male and female trees are used for the purpose. This is done by tapping the flowering shoots in case of male trees and the fruiting bunches for the female trees when drupes had not developed much. The male trees are usually tapped in winter while the female trees a few months later.

The sap is partially or highly fermented to prepare a crude liquor, known as toddy. Sugar in crystalline form is also prepared on boiling the sweet sap and the product may be brown or white according to processing method and temperature used.

The candy is another product made from the sap by boiling and crystallization, while molasses is prepared during crystallization. The seeds from immature fruits contain fleshy, translucent, mildly sweet, jelly-like endosperms which are much relished or also canned in syrup. The pulp, *i.e.*, liquid mesocarp of ripe fruits is used to make a number of fried products and a tough, leathery product which are liked by many.

Besides human consumable products, a number of other useful articles are also made from the tree. The fibre of the fruit is a raw material in many cottage industries, the leaves and midribs are used to make many household articles and the stem of the tree which has good timber value and durability is used as a post, foot-bridge etc. in the villages. Besides, the tree is planted for afforestation, as a wind-break and around large tanks, especially in West Bengal (Mazumdar, 1984, 2003). Medicinal value of the fruit is also known (Karunanayake *et al.*, 1984).

The palm is dioecious, tall, attaining a height of 20 metres and sometimes even upto 30 metres. The girth is 1-2 metres and the crown contains 30-40 leaves. The tree takes a long time to attain bearing maturity which may be 12-15 years. Male flowers are small and female flowers very large.

Flowering time of palmyra palm is from February to May and the fruits are harvested from July to October. Off-season fruits are also sometimes available in summer and scantily in winter. The fruit is spherical to oblate especially when 3 seeds are present and becomes curved when there are 2 seeds while more curved when contains 1 seed. The peel of the fruit is brown in colour with persistent calyx and the liquid mesocarp is rich yellow.

Palmyra palm is believed to be originated in India where it is grown in all tropical parts and especially in the coastal tropics. The estimated number of trees in India is 76,167,000 of which maximum is in Tamil Nadu which comprises 44.18 per cent (Anon., 1988).

Propagation of palmyra is done by seeds and these are not sown at more than 15 cm depth as deeper sowing causes rotting. Polyembryony in the seeds is also noticed but rarely.

Kandiah and Mahendran (1986) from Sri Lanka reported a new method for culturing seedlings and suggested technique on experimentation by which, the tubers could be grown in handy polythene bags. Veerasamy *et al.* (1995) had studied regeneration of shoots from split shoot apical meristem by exposing the meristem on removal of leaf-sheaths and foliage leaves and cutting them vertically into 2 or 4 equal parts.

Regular orcharding of palmyrah palm by application of inputs is uncommon under Indian conditions. However, in a trial done to study the effect of different organic manures, Velu (1989) observed that highest sap yield of 301.66 litres per palm could be obtained with application of 60 kg farmyard manure to a palm. The treatment also resulted in the longest possible tapping period and highest number of palms for tapping. The study undertaken by Arumugam *et al.* (1994) revealed that highest sugary sap yield of 113.82 litres per palm and quality with total soluble solids of 11.02 per cent by application of 50 kg farmyard manure alone per palm.

Hussain *et al.* (1992) had carried out a study on harvesting of juice as affected by different parameters such as climbing techniques, cutting methods, harvesting and juice collection methods under Bangladesh conditions and had ascertained techniques. The effect of application of a 20 mM solution of EDTA in increasing sap flow had been observed by Kandiah and Kokudathasan (1987).

Many insect pests which attack coconut trees are observed to attack palmyra also among which, *Rhynchophorus ferrugineus, Oryctes rhizocerus, Coccotrypses borassi* are more notorious. Among diseases, bud-rot due to *Phytophthora palmivora*, stem-bleeding by *Ceratostomella paradoxa*, leaf-splitting by *Exosporium palmivorum*, grey-blight by *Pestalotia palmarum* etc. are usually of much concern. Besides these, many other pests and diseases have also been on record (Anon., 1988).

Harvesting of fully ripe fruits of palmyra is only done and the normal tendency is to collect those which have naturally fallen. A palm may produce 150 or more fruits and this is much higher in Laccadive islands and Sri Lanka. No standard or named variety of palmyra is known.

The pulp of ripe fruits is reported to contain 77.2 per cent, 0.7 per cent, 0.2 per cent, 20.7 per cent and 0.7 per cent of moisture, protein, fat, carbohydrate and mineral matter respectively, besides 9 mg of calcium and 33 mg of phosphorus per 100 g of pulp (Anon., 1988). Mandal and Mazumdar (1995) harvested fully ripe fruits from several plantations in the western tableland of West Bengal and had made physico-chemical analyses of them. The average results revealed weight per fruit as 2.66 kg, specific gravity value as 0.88, stalk-stylar length as 15.1 cm and in the edible part of mesocarp, the total, reducing and non-reducing sugars respectively as 13.14 per cent, 6.47 per cent and 6.34 per cent while the total titratable acidity (equivalence of citric acid) was estimated as 0.29 per cent and the vitamin C content as 30 mg per 100 g.

Arseculeratne *et al.* (1982 a, 1982 b) had observed neurotoxic properties in the flour of palmyra. The existence of steroidal sapogenins (Flabelliferins) in the pulp had also been reported by Nikawela *et al.* (1988). The polysaccharide contents in the seeds and the possible effect of environment on those components had been put forward by Awal (1996) under Bangladesh conditions.

7

Carissa carandas Linn.

Common Names (English) : Carissa, Caraunda, Christ's
                                  thorn.

Indian Names   : Karamandaka, Karonda,
                                  Karanja, Karamcha, Karam-
                                  arda, Karavanda, Karekey,
                                  Vekao, Vakao, Kalakkai,
                                  Kalivi, Elimullu, Khirakoli,
                                  Karwanch etc.

Botanical Family   : Apocynaceae.

The berries of *karonda* are not taken as raw for very high acerbity in them and also for the astringency but they are of great demand for the preparation of a number of products. Jelly is by far, the most important product that can be named which is of excellent quality when made from these berries (Sinha Roy and Mazumdar, 1996). They are also dehydrated to make a nut-like product. In West Bengal, the immature berries are processed to make a cherry-like product (Mandal *et al.*, 1992). They are also of much use in making pickle, *chutney* and sour preparations.

The plants are grown to serve as highly protective hedge for the dense, bushy habit of their growth and especially for containing long thorns. They are planted in the wind-break rows as well. Besides, planting in slopy terrain to control soil erosion has also been a practice in the east Indian States.

The tree is evergreen, shrubby in growth and the height reaches more than 5 metres when unpruned. Leaves are glossy, dark green and ovate or elliptic in shape. In the leaf-axil and the apex, strong thorns are produced which may be forked. Flowers are white, borne in clusters of 2 or 3 and are of fragrance. Calyx is glandular, carpels are 2 and syncarpous.

The origin of the species has been stated to be India or Java (Ranjit Singh, 1969; Pareek and Sharma, 1993). In India, the tree is found to grow in all tropical and sub-tropical parts.

The tree does not grow in the Himalayan region but thrives well at some places in the Nilgiri hills (Sham Singh *et al.*, 1967). Soil is not so important factor for the trees. The experiment done by Gurbachan Singh *et al.* (1998) however, provided evidence that highly sodic soil is injurious to the plants. While testing performance of the trees at soil pH above 8.1, they noted that 80 per cent of the plants died within 3 months when the soil pH was 10.

Propagation of *C. carandas* is mostly done from seeds. Bankar (1987) studied influence of gibberellin on seed germination and seedling vigour of *C. carandas*. Seeds were treated by them with varying concentrations of GA_3 solution for 24 hours and the germination was found to be 67 per cent by treatment with 25 ppm of GA_3 solution as against 49.2 per cent in the control on the 58th day. Seed propagation in *C. carandas* has a difficulty in that, the growth of the seedlings is slow (Hayes, 1970). However, it was observed by Bankar (1987) that seedling growth could be enhanced by treating the seeds with GA_3 solutions. Misra and Jaiswal (1998) also studied the effect of GA_3 (0, 250,

500, 750 or 1000 ppm) applied 6 times at monthly interval on the growth of 9 months old seedlings and it was observed that seedling height and stem diameter increased with increasing concentrations. Improvement of germination of the seeds after longer storage period by treatment with the fungicide, Dithane M-49 (Mancozeb) had been an observation put forward by Singh, J. and Tiwari (1998).

The cuttings do not form roots easily in *C. carissa*. But treatments with auxins in suitable concentrations have been reported to promote rooting (Bandopadhyay *et al.*, 1983; Sanjay Tyagi *et al.*, 1999). The effect of application of varying concentrations of IBA and NAA on stool layering had been studied by Misra and Jaiswal (1993) and success had been observed by them with best result at 10,000 ppm of IBA. Misra and Singh (1990) reported success in air-layering by treatments with IBA and NAA and obtained maximum rooting when 5000 ppm of IBA was treated. They concluded that IBA had been more effective than NAA to promote rooting in air-layers. In *C. carandas*, grafting is also done to a limited extent by inarching (Sham Singh *et al.*, 1967).

In a study done in determining the laminar nitrogen, phosphorus and potash content of the leaves of *C. carandas* grown in West Bengal at monthly interval from December to May, De *et al.* (1996) observed that sharp increase had occurred upto the month of March for nitrogen and potassium and upto the month of February for the phosphorus. Thereafter, the components had declined consistently and the decrease had been sharp from April to May.

Trees of *C. carandas* start flowering from February and fully mature fruits are obtained from August to September or even October. Harvesting of the unripe fruits is started from May.

No true variety of the species is known but two types of fruits are observed, one having dark purple and almost black coloured fruits and those of the other of creamy yellow colour with attractive pink blush. Mehra and Arora (1982) stated diversity of the tree characters in the species grown in Mount Abu, Khandala and other parts of north-western India. According to Karale *et al.* (1989), some promising types have been identified in Maharashtra.

The fresh fruits of *C. carandas* have been reported to contain 91 per cent moisture, 1.1 per cent protein, 2.9 per cent fat, 0.6 per cent minerals, 1.5 per cent fibre, 2.9 per cent carbohydrate, 0.021 per cent calcium and 0.028 per cent phosphorus (Gopalan *et al.*, 1993). The composition of the dried berries has also been on record (Sham Singh *et al.*, 1967; Gopalan *et al.*, 1993). Singh, A.K. and Singh, P. (1998) chemically analyzed red-green and white-yellow berries and opined that the latter berries which were less sour could be more suitable for making jam, jelly, *chutney,* pickle and preserves. Significantly high pectin content and of high jelly-grade in the berries (determined by crude method) grown in the southern part of West Bengal has been reported by Sinha Roy and Mazumdar (1996) and Majumder and Mazumdar (2001 b). A number of workers have reported oil and other compounds also from the leaves, flowers and fruits of *C. carandas* (Chandra, 1985; Naim *et al.*, 1988; Sekar and Francies, 1998).

The changes taking place during development of the fruits had been a study undertaken by Uthaiah (1988) in Karnataka. Sinha Roy (1988) sprayed *C. carandas* trees grown in West Bengal with 0.5 per cent, 1 per cent or 1.5 per cent solutions of urea, potassium chloride or magnesium sulphate, or 0.1 per cent, 0.25 per cent or 0.4 per cent solutions of copper sulphate or zinc sulphate, or 0.05 per cent, 0.1 per cent or 0.15 per cent solutions of sodium tetra-borate or, 20, 60 or 100 ppm solutions of IBA, or 10, 40 or 70 ppm solutions of GA_3 or ethephon, or 5, 15 or 25 ppm solutions of kinetin after 7-10 days of fruit set. The berries sprayed with the chemicals or with water (control) were analyzed for calcium pectate and jelly-grade of the pectin (by crude method) at four equally spaced stages of their development till attainment of harvest-maturity. Among other observations, it has been noted that in heightening the pectin content at different maturities, nutrient salts were more effective with best effect induced by 1.5 per cent KCl solution. Among growth regulators, 70 ppm of ethephon produced best result. Jelly-grade of the pectin had not been significantly enhanced by the treatments. The water-soluble, oxalate-soluble, acid-soluble and alcohol-soluble pectic substances were also determined for the berries. In West Bengal, the berries of *C. carandas* are processed by imparting colour and sweetness so as to look like red-coloured cherries and these are used as an ingredient in making various sweatmeats and also in confectioneries. Slightly immature berries which have not attained full maturity are only utilized for the purpose. Mandal (1997) sprayed the trees with 0.5 per cent, 1 per cent or 1.5 per cent solutions of urea, potassium chloride or zinc sulphate, or with IBA (20, 60 or 100 ppm), or GA_3 or ethephon (10, 40 or 70 ppm) after 10 days of fruit set. The sugars, acidity, size and shape of the berries were determined with progress of development. It was observed that the ideal maturity for making cherry could be hastened and thereby, extended for at least 10 days more by some treatments, notably 1.5 per cent solution of urea or zinc sulphate or 100 ppm solution of IBA.

8

Chrysophyllum cainito Linn.

Common Names (English) : Butter lime, Star-apple, West Indian star-apple.

Indian Name : (Not known)

Botanical Family : Sapotaceae.

Star-apple is used as an ornamental tree for its leaves of striking appearance but the fruits have also edible value (Sham Singh *et al.*, 1967).

The tree attains a height of 10 metres or more. The leaves are oval to oblong in shape, 8-12 cm long, deep green in colour and glabrous above but under-surface is covered with yellowish brown or coppery red pubescence. Fruit is globular, apple-shaped, 5-10 cm in diameter, smooth with either green or purple surface colour. Pulp is soft, white and sweet when ripe. On transverse section, the fruit is star-shaped as the location of the brown coloured seeds and the seed-cores present such an appearance.

The origin of the tree is central America and West Indies and in India, it grows as wild in warmer parts, especially around Baroda and Mumbai (Ranjit Singh, 1969). On the Nilgiris, the tree thrives upto 1,067 metres (Sham Singh *et al.,* 1967).

Propagation of star-apple normally takes place by seeds. In some countries, budding is also done. Anaya Hernandez and Vega Cuen (1992) in Mexico extracted seeds from small, medium-sized or large fruits to study germination behaviour. Seeds were sown in soil or river sand at 3 or 5 cm depth and maintained in sun or shade. Germination percentage was found to be highest (78.5 per cent) in the seeds obtained from medium-sized and large fruits that were sown at 3 cm depth in river sand and maintained in shade. Most of the seedlings (97.9 per cent) had survived on transplanting in soil.

Knight and Campbell (1993) studied pollination requirement of the tree in Florida, USA. Harvesting of the fruits should be done when fully ripe, *i.e.,* from end of January to March under Indian and especially south Indian conditions. A tree may yield upto 70 kg fruits.

9

Dillenia indica Linn.
(*Dillenia speciosa* Thunb.)

Common Name (English) : Dillenia.
Indian Names : Chalta, Karambel, Chalita, Peddakalinga, Uva, Punna, Autenga, Betta Kanigalao etc.
Botanical Family : Dilleniaceae.

Dillenia is a dual purpose tree in that, it yields edible fruits as well as good quality timber. The fruit is botanically a pseudocarp and the tightly attached fleshy acerescent sepals form as the edible parts.

The sepals are highly acidic and hence, are not possible to consume as fresh. But a number of products can be made from these sepals. In West Bengal, it is a very popular fruit to prepare *chutneys,* sour preparations and pickles. The wood is used for making packing cases (Pareek and Sharma, 1993). The tree and the leaves are graceful in appearance for which, it is also planted along avenues.

The origin of the tree is India (Pareek and Sharma, 1993) and according to Ranjit Singh (1969), in east India.

The tree is found to grow in high rainfall area and especially in West Bengal, Assam and the eastern States. In south India, it is largely grown in the coastal areas. The tree is a common occurrence in evergreen forests of the sub-Himalayan tract (Pareek and Sharma, 1993).

The dillenia tree is grown as wild and when planted, no cultural or manurial treatments are adopted. De *et al.* (1996) sampled leaves borne by trees grown in the South 24-Paraganas district of West Bengal and analyzed nitrogen, phosphorous and potash contents at monthly interval from the months of December to May. All the constituents were observed to increase steadily although at different degrees and the rate of increase was greater from the months of March to May for nitrogen and phosphorous contents.

According to Gopalan *et al.* (1993), the fruits contain 82.3 per cent moisture, 0.8 per cent protein, 0.2 per cent fat, 0.8 per cent minerals, 2.5 per cent fibre, 13.4 per cent carbohydrate, 0.016 per cent calcium and 0.026 per cent phosphorus. Physico-chemical analyses, done by Mandal and Mazumdar (1995) with fully mature fruits harvested from trees grown in the southern part of West Bengal revealed average weight per fruit as 592 g, specific gravity as 0.98, stalk-stylar length as 8.8 cm, total sugars as 0.61 per cent, reducing sugars as 0.43 per cent, non-reducing sugars as 0.17 per cent, total titratable acidity (equivalence of citric acid) as 0.69 per cent and the vitamin C as of 20 mg per 100 g of pulp.

Physico-chemical changes during development of dillenia fruits under Assam conditions have been done by Neog and Mohan (1993). It has been reported that the fruits took about 160 days to reach harvest-maturity from fruit set when they were straw-yellow in colour with a greenish tinge. The percentage of TSS, titratable acidity, reducing sugars, total sugars and the vitamin C content of the fruits at harvest-maturity have also been stated by them. Mandal and Mazumdar (1998) sampled dillenia fruits at four equally spaced stages of development from fruit set till attainment of harvest-maturity and had analyzed those for weight, specific gravity, stalk-stylar length, cross-sectional diameter, dry matter, total, reducing and non-reducing sugars, titratable acidity and the vitamin C content. In general, it was observed that gain in fruit weight, length, diameter, dry matter, total and reducing sugars had been higher at the later part of development. Other components of the fruits have also been studied (Mandal, 1997).

10

Euphoria longan (Lour.) Steud (*Dimocarpus longan* Lour.) [*Nephelium longana* (Lamk.) Camb.] (*Euphoria longana* Lamk.)

Common Names (English) : Longan, Lungan.

Indian Names : Anshphal, Tokra, Puvatti, Mulay, Poripuna, Oomb, Shempuvan, Malakota, Mulei, Mora, Pasakota, Kattu puvam, Naglichi, Murale etc.

Botanical Family : Sapindaceae.

From the horticultural point of view, the longan is not as important as litchi (*Litchi chinensis*) which is the other species of the same family. Nevertheless, the fruits of longan are liked by many in fresh form. Canning of the fruits is also done especially in the south-east Asian countries. In

Thailand, drying of longan is very common which is exported to other countries as well.

The tree of longan attains a height of 15 metres or more. Leaves are glabrous. Inflorescence is terminal, hairy and the flowers are yellowish to brown. Flowers are unisexual but the male flowers have non-functional stigmas. In female flowers, the ovary is conspicuous and consists of two carpels. Fruits are spherical or oblong with thin peel and the tubercles of the skin are not conspicuous. The edible part is a fleshy aril, translucent, sweet and has flavour. Seeds are brownish to black, spherical or ovoid and separate easily from the aril.

The origin of the species has been stated to be India-Burma (Ranjit Singh, 1969) but may be southern China or the region between India and Myanmar. Ke Guan Wu *et al.* (1994) studied morphology of pollen grains in 14 cultivars of longan and also wild species from 5 regions of China.

From differences in the pollen grain veins observed by them and from analysis of botanical geography and evolution, they held that Yunnan might be the primary centre of origin and Guangdong, Guangxi and Hainan provinces as the secondary centre of origin for the species.

In India, the tree grows in all tropical parts having well-distributed rainfall and mild summer and winter. Dry and hot weather in summer result in heavy drop of fruits and also cause cracking of them. Soil is not an important factor but slightly acidic soil is considered superior. Inoculation of the soil before planting by mixing a handful soil taken from the soil of a mature tree should be done.

Propagation of *E. longan* is commonly done with seeds but true-to-type trees are not produced in the seed-propagated trees. Seeds loose viability with storage and hence, freshly extracted seeds should only be used. Xia *et al.* (1992 a) studied that optimum temperature for germination was 25°C. They stated that after moist storage in perlite at 20 per cent moisture in sealed polythene bags for 15 days or soaking in 250 ppm of GA_3 solution for 24 hours, the longan seeds reached 100 per cent germination at 30°C after 5 days. It was observed that the seeds almost lost viability when their moisture content decreased to around 29 per cent after slow drying with silica gel. The said workers (Xia *et al.*, 1992 b) had provided evidence that storage of longan seeds in perlite with 20 per cent moisture was preferable to storage in polythene bags. The chemicals which controlled fungal infection and spontaneous germination in storage have also been named by them. Other aspects of storage of the seeds have been studied by these workers.

Root formation in *E. longan* by stem cuttings is difficult but provision of bottom heat may be helpful to achieve some success as in case of litchi. In Thailand, air-layering is also done which gives high success and the layers become ready within four months.

Grafting and budding are largely done in China following a number of methods. The temperature of 20°-30°C is considered congenial for grafting.

Chen Jing Ying *et al.* (1999) described procedure for micro-propagation of *E. longan* using lateral and terminal shots. According to them, culture in liquid medium with 0.01 mg of 6-BA and 1 mg of IBA per litre was best for rooting. While studying *in vitro* culture of 0.5 cm shoot tip explants on MS and N_6 media, Wang and He (2000) stated that the combination of 0.3 mg of BA along with 0.2 mg of IAA and 0.5 mg of GA_3 per litre was most satisfactory among other treatments.

Pruning is a very important operation for the tree to encourage new shoots which actually produce flowers. In Thailand, it is a regular practice and is done even by machines. Analysis of the leaves in respect of nitrogen, phosphorus and potash content had been done by De *et al.* (1996) at monthly interval from winter to summer taking samples from trees growing in the southern part of West

Bengal. It has been observed that the nitrogen content had progressively increased upto the month of April and the other two elements upto the month of March. Subsequently, the elements tended to decline in the summer season when the fruits had started growing.

Zhuang *et al.* (1995) had made a study on the ranges in leaf contents of 10 elements in Tongan, Fujian in China. They held that the elements were affected by locality and year and the co-efficients of variation were low for nitrogen and successively higher for K, Mg, P and Ca. Besides, the variation was found to be greater for micro-elements than the macro-elements. In Thailand, spraying the trees with solutions of nutrient salts is also a common practice.

Promotion of flowering by ringing of shoots had been studied by Wu Ding Yag *et al.* (2000) in China who put forward that the practice significantly increased carbohydrate, starch and the C/N ratio in leaves and thereby, enhanced flowering and fruiting of the trees to a large extent.

There have been a large number of varieties of *E. longan*. In southern China alone, at least 300 varieties exist and about 50 varieties are known to be in cultivation. Some famous varieties which grow there are Fu Yan, Wu Long Ling, Wu Yuan, Shi Xia, Chuliang etc. Among the notable varieties of Thailand, mention may be made to those like Biew Kew, Dang, Daw, Haew, Chompu etc. Huang Jin Song *et al.* (1996) reported a large-fruited, late ripening variety *viz.*, Baoyuan in the Fujian province of China and Yang Win Huan (1996) discovered a superior chance seedling in that province. Chao Cheng Nan *et al.* (1997) described fruit characteristics of 42 local varieties in Taiwan. In Fujian, Xu Xiu Dan *et al.* (1999) reported an extremely late maturing variety, *viz.*, Lidongben and Chen *et al.* (2000) had mentioned about the cultivar, Qingshan which was evolved as a chance seedling. Any standard variety is not grown in India but a large number of variations are found in the tree and fruit characters of the trees grown in this country.

Analysis of the fruits harvested in the southern parts of West Bengal done by Mandal and Mazumdar (1995) revealed that average weight of fruit was 6.81 g with a specific gravity of 1.04 and the length per fruit was measured as 2.38 cm. The percentage of titratable acidity (equivalence of citric acid) in the aril was determined as 0.13 and the vitamin C as 10 mg per 100 g. The sugar: acidity of ratio of the aril was determined as 52.23. Pectin content as calcium pectate in the aril of the fruits harvested from same place could not be observed to be more than 1 per cent (Majumder and Mazumdar, 2001 b). The volatile compounds in the aril have been reported by Wong *et al.* (1996) in Malaysia and Chang Ching Hui *et al.* (1998) in Taiwan.

Bhattacharyya *et al.* (1990, 1992) and Mandal and Mazumdar (1998) had determined the weight, specific gravity, length and diameter of the fruits and the contents of dry matter, total, reducing and non-reducing sugars, titratable acidity and the vitamin C in the aril at four stages of longan fruits during their development till attainment of harvest-maturity condition under the conditions of West Bengal. The general trend of results showed that increase in fruit weight had been higher at the early part of growth while the specific gravity had made much increase during the middle part. Increase in sugars and decrease in acidity had been observed to be more sharp at the advanced maturity condition of the fruits.

11

Flacourtia indica (Burm. f.) Merr. (*Gmelina indica* Burm. f.) (*Flacourtia ramontchi* L' Hérit) (*Flacourtia sepiara* Roxb.)

Common Names (English) : Governor's plum, Madaga-
scar plum, Batoko plum,
Paniala plum, Ramontchi etc.

Indian Names : Paniala, Bainchi, Kantai,
Talispatra, Kandregu,
Bilangra, Kanju, Kankod,
Katukala, Bhekal, Tambat,
Saralu, Talisam, Goraji,
Ramontchi, Hattari mullu,
Binja, Kaker, Vayankatha,
Hunmunki etc.

Botanical Family : Flacourtiaceae.

Both the species of *Flacourtia* grow as wild trees or the
seedlings are also planted to erect protective hedges. These

bramble fruits are edible as fresh although some of them taste very sour.

Hayes (1970) quoted that the fruits are greatly improved by rolling between the palms before eating. However, both the sweet and the sour fruits serve as excellent materials for preparation of jelly, jam, pickles, tarts, preserves etc.

The trees of both the species are thorny and shrubby in growth habit. The height may go upto 8 metres or more. The branches sparsely develop thorns at the leaf-axils which are pointed very long and may be 2 cm. The trees produce dense foliage. The leaves are shiny, deep green, oblong-obovate, measure 5-7 cm in length and have dentate-serrate edges.

The fruits are small and 2-3 cm in diameter and the colour is red or purple and sometimes dark purple. Many variations are however, noted in the plants produced from seeds. Thus, in some seedling plants fruits are sweet, sometimes appreciably sweet while the fruits borne by other plants taste sour and sometimes very sour. The seeds are small, thin and 12-14 in number.

The origin of the species is believed to be Malagasy or India. In India, the trees of both the species are found to grow in West Bengal, Assam, the eastern States, Maharashtra, Bihar, eastern U. P. and at the foot of the Himalayas.

The trees are found to grow in inferior soil having poor fertility but do not grow well in places where water stagnates. Tropical and moist sub-tropical climate are suitable for both the species and according to Hayes (1970), the trees are drought resistant.

Both the species are propagated easily from seeds but vegetative propagation needs encouragement for the plants which produce sweet fruits. Bora and Das (1998) from Jorhat, Assam reported propagation by cuttings with use of suitable concentrations of IBA. Similar viewpoint has also been held by Jayant Kumar and Parmar (2000) from Himachal Pradesh. Propagation may be done by inarching or budding as well, using seedlings of the same species as rootstocks (Hayes, 1970).

Fruitfulness is a problem in *Flacourtia* as the trees are dioecious. Hence, several trees should be planted together for cross-pollination. No culture is required by the trees but Hayes (1970) commented upon that irrigation during fruit development may be given, if there is no rain.

The fruits have been reported to contain 67.8 per cent moisture, 1.7 per cent protein, 1.8 per cent fat, 1.3 per cent minerals, 4.7 per cent fibre, 22.7 per cent carbohydrate, 0.1 per cent calcium and 0.1 per cent phosphorus (Gopalan *et at.*, 1993). Bhaumik *et at.* (1987) isolated flacourtin, which is a phenolic glucoside ester from the bark of the trees and provided structural elucidation of the compound.

12

Grewia subinaequalis DC.
(*Grewia asiatica* Masters)

Common Name (English)	: (No specific name).
Indian Names	: Purusha, Phalsa, Dhamin, Dhamani, Polisha, Jana, Pharasakoli, Nalajana, Shukri, Tadasala, Pharnu, Chadicha, Phutiki, Buttiudippe, Tadachi, Dowaniya etc.
Botanical Family	: Tiliaceae.

Phalsa is one of the hardiest, drought-resistant and quick-growing fruit crops of India. It is grown as wild but is also planted (i) in homestead land for fruits, (ii) for bushy hedge, (iii) for wind-break, (iv) in social and agro-forests for the fruits as well as the pruned parts for use as fuelwood, (v) for productive utilization of waste land, (vi) to grow as filler plants in mango and other orchards and (vii) in hilly area to control soil erosion.

The ripe fruits are relished afresh but are also utilized in making juice, syrup, drinks, sherbets etc. and these make great appeal as the ripe fruits are obtained in summer season. Good quality squash and jam are sometimes made from the fruits, usually by blending with other fruits. In many parts of north India, strong and durable baskets are also made from the pruned branches.

The tree bears flowers on current season's growth with 2-9 peduncles and each peduncle bears 3-6 pedicellate flowers (Pathak and Pathak, 1993). Petals of the flowers are thickened and have glandular claws. Fruits are smooth and drupaceous.

The origin of the species has been stated to be Baroda region in India (Ranjit Singh, 1969). It is found to be grown in all parts of the country and largely in the northern and the central regions.

Phalsa grows both in tropical and sub-tropical climate but productivity and sweetness of the fruits are much greater in the sub-tropics. In sub-tropical climate, it is deciduous, but usually not so in the tropical situation. Low winter temperature is helpful for flowering. Summer temperature upto 45°C is tolerated well by the plants. Soil is not an important factor for *phalsa* and inferior type of soil is generally utilized for the fruit crop.

Seed propagation of *phalsa* is very common in India. But the seeds lose viability greatly within 3 months when stored in ambient condition and hence, fresh seeds from fully ripe fruits should only be sown.

Experiments done by Darbara Singh and Jindal (1979) also observed that germination percentage was highest (76 per cent) and germination periods was shortest (16 days) when seeds were sown immediately after extraction than when the seeds were stored. Formation of adventitious roots, *i.e.,* cuttings and layerings are however, difficult in *phalsa* but may be successful by the use of auxins. Vegetative propagation may be done by grafting, preferably by pre-defoliated scion sheets and adopting veneer or whip grafting method.

No manure or fertilizer is applied when of *phalsa* is grown but productivity and quality of the fruits are found to be tremendously improved by such application. Singh, R.P. and Gaur (1989) made a 3-year fertilizer trial in Kanpur using 28 combinations involving different levels of urea, single superphosphate and muriate of potash. It was observed by them that the fruit yield was maximum when 100 g N, 40 g P_2O_5 and 40 g K_2O were applied per plant. Besides, growth of the plants was found to be much affected by the fertilizers. Analysis of the leaves in respect of N, P, K, Ca, Mg, S, Fe, Mn, Zn and Cu contents have also been done by Bhargava *et al.* (1997) for nutrient diagnosis.

The most important operation in the cropping of *phalsa* is pruning. This should be done annually when the leaves are shed. Very drastic type of pruning is though practised by some growers in north India but it is perhaps unnecessary (Hayes, 1970). The works by Rao and Reddy (1989) produced evidence that highest fruit yield was obtained when the bushes were pruned at 75 cm above the ground level as compared to 15, 30, 45 or 60 cm height or when no pruning was done. Whatever may be the amount of pruning, the operation is essential for augmenting growth and productiveness (Hayes, 1970).

Phalsa is a self-pollinated tree and controlled pollination greatly increases fruit set (Pathak and Pathak, 1993). Parmar (1976) reported that self-pollination by hand, natural selfing, open pollination and hand-crossing respectively resulted in 59.9, 51.1, 71.6 and 81.3 per cent fruit set. They added that honey bees were the most important pollinating insects.

The yield of *phalsa* may go upto 10 kg per plant and the yield may be raised by suitable pruning and controlling the bark-eating caterpillar which is sometimes very serious.

Any established variety of *phalsa* is not there. Some types which have umbrella-like plants are called *Chhatri* as compared to those which are bushy in appearance. The plants of some type are found to be tall and some are found to be dwarf. A type, *Sharbati* is known in north India, which is believed to be of better quality. Nehra *et al.* (1985) presented tabulation data on the plant, flower and fruit characteristics of the tall and dwarf forms of *phalsa*.

The chemical composition of *phalsa* fruits has been done by a number of workers. Hasnain and Ali (1988) evaluated the crude protein percentage and amino acids in *phalsa* fruits grown in Pakistan and their study revealed the presence of two major proteins. The composition of *phalsa* fruits presented by Gopalan *et al.* (1993) states moisture 80.8 per cent, protein 1.3 per cent, fat 0.9 per cent, minerals 1.1 per cent, fibre 1.2 per cent, carbohydrate 14.7 per cent, calcium 0.129 per cent, phosphorus 0.39 per cent and iron 0.003 per cent.

Mandal and Mazumdar (1997) analyzed *phalsa* fruits grown in the southern part of West Bengal and presented results for weight, specific gravity, length and diameter per fruit as well as the percentage of dry matter, titratable acidity (equivalence of citric acid), total, reducing and non-reducing sugars and the vitamin C of the fruits. The study undertaken by Majumder and Mazumdar (2001 b) revealed a pectin percentage of about 1 per cent as calcium pectate in the fruits grown in the southern part of West Bengal. Moti Singh *et al.* (1986) reported that fruit set, yield and fruit quality as TSS were best when the bushes were sprayed with solutions of GA at 60 ppm with ethephon at 1000 ppm concentrations. Increased percentage of fruit set has also been observed by Zora Singh *et al.* (2000) by spraying gibberellin solutions.

13

Limonia acidissima Linn.
[*Feronia limonia* (L.) Swingle]
(*Feronia elephantum* Correa)

Common Names (English) : Elephant apple, Wood apple, Indian wood apple.

Indian Names : Kavittha, Kaith, Kath bael, Kapith, Kaitha, Kavit, Bilin, Kotha, Vila, Kothum, Vilanga, Vilati, Vila, Bhanka, Velamarum, Velaga, Diwul etc.

Botanical Family : Rutaceae.

The tree of elephant apple grows as wild in the plain lands all over India. But it is also planted along avenues, in forests and in household compounds.

As a fresh fruit, elephant apple is not very popular but a number of products are made from the fruits. Hayes (1970) stated that a stiff jelly could be made from these fruits but the flavour is somewhat harsh and hence, is seldom used. According to him, the fruits can be mixed with guava to make good quality jelly. According to Poi and Mazumdar (1989), the *kath bael* fruit is much in demand in West Bengal to make *chutney*. This is done more commonly by mixing with molasses than sugar, so as to make a paste. In some parts of India, ripe fruits are utilized to make drinks.

The tree is also known (Purohit *et al.*, 1996) to have high medicinal value, astringent properties and has beneficial role in cardio-vascular system. Bandara *et al.* (1989) in Sri Lanka made suitable extracts from stem bark of the tree and found the same to have high insecticidal properties. Zarga (1986) from Jordan reported 3 indole alkaloids from woody stem of the tree. According to him, the stem powder of the tree is also used as a cosmetic in Burma. Natural products having medicinal, insecticidal, anti-fungal and other properties have also been reported by other workers (Ghosh *et al.*, 1989; Macleod *et al.*, 1989; Parthasarathi Ghosh *et al.*, 1991).

Pareek and Sharma (1993) stated that *L. acidissima* induced early flowering in *Citrus* when used as a rootstock.

The tree of the species is tall and attains a height of 11-12 metres. Fruit is amphisarca, has hard rind and edible parts constitute the succeulent placenta and the inner pericarp.

The tree is a native of India (Ranjit Singh, 1969) and is distributed all over the country but prefers a dry, sub-tropical climate (Pareek and Sharma, 1993).

Propagation is normally done by seeds although root cuttings or layers are also successful (Hayes, 1970). Poi and Mazumdar (1989) experimented with leafless semi-hardwood and leafy softwood stem cuttings and treated the basal parts with 50, 100 or 200 ppm of IAA, IBA or NAA solutions. The auxin treated or untreated cuttings were maintained in intermittent mist and only the semi-hardwood cuttings treated with IAA (100 ppm) or NAA (50 or 100 ppm) were found to produce roots. The survival percentage after six months was 26.6 with IAA and 62.2 and 40 respectively with 50 and 100 ppm of NAA. Fourie (1988) claimed success of *in vitro* grafting of apices of elephant apple on troyer citrange rootstocks aiming at producing virus-free plants.

A method for *in vitro* proliferation of shoots has been presented by Handique and Bhattacharjee (2000), using B-5 medium supplemented with BA and kinetin. Purohit *et al.* (1996) studied biochemical changes occurred during process of shoot regeneration in callus cultures of *L. acidissima*. They held that the process of de-differentiation and re-differentiation had high metabolic activity.

With study done in aiming at *in vitro* organogenesis of cotyledon explants, Hossain *et al.* (1994 a) used MS medium containing BA and kinetin. Adventitious buds formed on the cotyledons were found to have developed into shoots that rooted in half-strength MS medium containing 0.57 micro-M of IAA and 0.49 micro-M of IBA. Plantlets were observed to have successfully established in soil.

De *et al.* (1996) analyzed nitrogen, phosphorous and potash content of leaves explanted peripherally from trees grown in the central part of West Bengal. The sampling done at monthly interval from the months of December to May made it apparent that the constituents made increase throughout and at a higher rate during the summer season.

Pollination of *L. acidissima* had been a study undertaken by Baskaran and Bal (1998) in Tamil Nadu. They observed that the tree was entomophilous and the dominant species was *Apis cerana indica* which caused about 92 per cent pollination. Pollination activity had been noted to be maximum from 10 to 11 hours.

As regards fruit composition, the report by Gopalan *et al.* (1993) states, moisture-64.2 per cent, protein-7.1 per cent, fat-3.7 per cent, minerals-1.9, fibre-5.0 per cent, carbohydrates-18.1 per cent, calcium-0.13 per cent, phosphorous-0.11 per cent and iron-0.048 per cent.

14

Malpighia glabra Linn. and *Malpighia punicifolia* Linn.

Common Names (English) : Barbados cherry (for *M. glabra* L.) and West Indian cherry, Acerola etc. (for *M. punicifolia* L.).

Indian Names : Nelli, Simeyara nelli, Vellari etc.

Botanical Family : Malpighiaceae.

Fruits of these species have their pride in having highest vitamin C content among all human consumable fruits so far known. Besides, jellies, jams and sherbets are also prepared from these fruits with or without blending of other fruits. The shrubs are planted in gardens also for their decorative value.

There has been some confusion over the botanical nomenclature of the Barbados or West Indian cherry. The shrubs reach a height of upto 3 metres in *M. glabra* and much more in *M. punicifolia* and are spreading in habit.

The branches are slender, leaves are ovate-shaped, shiny and 2-6 cm long. Flowers are axillary umbellate, white to rosy in colour and have 5 petals. Fruits are small like cherry, red in colour with thin skin and have 3 crested seeds. Morphology and anatomy of seeds have been studied by Nacif *et al.* (1996).

The origin of these species is stated to be south America (Ranjit Singh, 1969), or West Indies and in the area from the northern part of south America to southern Texas in the USA. These tropical species have been introduced into India and now found to grow well in eastern U.P., West Bengal, eastern India and parts of south India.

Propagation of the species is done by seeds but great variations in the seedlings of *M. glabra* have been reported by Carvalho (1998) in Brazil. However, stem cuttings and layerings are also successful with or without use of auxin. Gonzaga *et al.* (1996) from Brazil reported success with splice grafting (40 per cent), patch budding (86.7 per cent) and also cleft grafting in *M. glabra*. Leaf explants of the same species were cultured by Alloupa and Campos (1999) in Brazil and noted that callus formation was best in MS medium, supplemented with suitable concentrations of kinetin and 2, 4-D.

Brasil *et al.* (1999) in Brazil made a fertilizer trial of *M. glabra* seedlings in greenhouse using 4 levels of urea, 4 levels of potassium chloride with or without application of 2 levels of limestone. Various responses were found on growth of the plants but potassium and limestone had little effect.

Informations on pollination requirements in respect of pollinators, pollenizers and the number of plants required have been provided by Knight and Campbell (1993) in *M. punicifolia*. Results of pollination under different treatments have also been reported by Magalhaes *et al.* (1999) in the same species.

According to Gomez *et al.* (1999), the shrubs of *Malpighia* species yield 6 to 60 tonnes of fruits per hectare. Mezuita and Vigoa (2000) stated annual yield of the two species as 9 to 13 tonnes per hectare in Chile. Increase in yield by application of pacrobutrazol in suitable proportion has been reported by Michelini and Chinnery (1988).

Visentainer *et al.* (1997) reported average content of vitamin C in the fruits of M. *glabra* as 1.79 g per 100 g of pulp. According to them, the fruits contain 89.09 per cent moisture, 7.35 per cent carbohydrate, 0.17 per cent fibre, 0.20 per cent lipids and minerals. Gomez *et al.* (1999) stated that the fresh fruits of *Malpigha* species contained 1 to 3 g of vitamin C per 100 g of pulp. Upto 4 g of vitamin C per 100 g of fruits had also been estimated by Attri and Singh (1997) in fully ripe fruits of M. *glabra* grown in the Andaman and Nicobar islands.

The vitamin C and other components in the fruits of the two species during their development have also been reported by a number of workers (Guadarrama, 1984; Cruz *et al.*, 1995; Attri and Singh, 1997; Vendramini and Trugo, 2000) and post-harvest storage of M. *glabra* fruits by Chou Hui Na *et al.* (2000).

15

Monstera deliciosa Liebm.
(*Monstera lennea* Koch.)

Common Name (English) : Ceriman.
Indian Name : Amar phal.
Botanical Family : Araceae.

This is primarily an ornamental vine but the fruits have also edible value. Under Indian condition, the climber is grown in gardens, in houses or household compounds for decorative purpose and not much grown for fruits.

The plants are vigorous, straggling and produce large number of broad, long-petioled, leathery leaves, which are perforated and scalloped. The vine clings to its support on rough surface by heavy aerial roots. The spathe in flowers is white and waxy and it covers a green spadix.

Fruits which develop from the spadix are large and like cone in shape. They resemble large ears of maize and take long time to mature which may be 8-10 months or even more after set. On maturity, the green colour of the fruits turns yellow and the scales which cover their surface become loosened and fall off. While the fruit is cut, the stem is placed in a pot containing water and when the lower scales fall off, it is ready for eating. Ripening of the fruit starts from base upwards and not all at a time. Accordingly, the part of the fruit which has ripened may be eaten and remaining part may be stored in cold for later use (Sham Singh *et al.*, 1967).

The pulp of the fruit is soft and has a pineapple-banana flavour. But it does not appeal many people perhaps because of the presence of calcium oxalate spicules or crystals which cause irritation to the tongue or throat of them, although not to all.

Apart from horticultural aspect, *M. deliciosa* has much been studied for botanical and for plant physiological and biochemical interest as well (Genua and Hillson, 1985; Zeier and Schreiber, 1998). Sanabria *et al.* (1999) studied morphology of the leaves with the idea to develop an identification system for ornamental species of *Monstera*. Floral anatomy of *M. deliciosa* has been studied by Barare and Cheretien (1985) in Canada.

The plant is originated in Mexico-Guatemala (Ranjit Singh, 1969) and is grown in all tropical parts of India.

In highly humid conditions, the vine grows vigorously and produces very large leaves. Mortensen (1991) in Norway grew plants in pots at temperatures ranging from 15°-33°C and noted that 24°-27°C as the optimum for temperature.

For propagation, long stem cuttings containing two or more buds are used. Seeds are also produced by the plant and may be used for propagation. Li Yu Qiao and Zhou CuiPing (1997) from China provided detailed method on the most suitable culture media for *in vitro* culture of *M. deliciosa* for micro-propagation.

The role of the type of fertilizers on *M. deliciosa* has been studied by Moschine *et al.* (1985) by using different nutrient salts in ion-exchange resin and it has been stated that the requirement of nutrients is fairly high for the vine. Nitrogen and potassium nutrition of the plant has also been studied by Khattab *et al.* (1987) in Egypt.

The effect of three substrates and three irrigation systems in *M. deliciosa* has been tested by Beel (1988) in Belgium. Comparative growth and tissue analyses in *M. deliciosa* were carried out by Yeh DerMing (1998) in Taiwan by growing the plants in a sub-irrigation system with three nutrient concentrations, in tap-water and ebb-and-flow systems and they had worked out method for production of high quality plants. Schelstraete and Beel (1985) from Italy reported pot-cultural method for production of good quality plants of *M. deliciosa*.

16

Phoenix sylvestris (Linn.) Roxb. (*Elate sylvestris* Linn.)

Common Names (English) : Wild date, Date-sugar palm etc.

Indian Names : Kharjuri, Jangi khejur, Kharak, Ita, Sendhi, Icham, Biochand, Itchalu, Kharjur, Pindakh-arjura, Kojari, Indi etc.

Botanical Family : Palmae.

Although wild date palm does not produce fruits of such quality as are comparable to those produced by the cultivated date (*P. dactylifera*), the tree of this species has other utilitarian value. The sap obtained by tapping the trees could be mentioned first which is used in making jaggery, sugar and other products or is also used as a drink, with or without fermentation. The fruits have edible value and are also used to make jelly or jam besides dehydration. Medicinal properties of the fruits is also known (Bajpayee, 1997).

The seeds of this palm are sometimes used for extraction of oil (Kapoor *et al.*, 1990; Bhakare *et al.*, 1993). In some parts of India, the seeds are also taken with betel leaf as substitute of areca nut. The trunk of the tree is used as a post in villages and the leaves are used in handicrafts (Mazumdar, 1984; Kamaluddin *et al.*, 1998).

The tree has a large crown and attains a height of 10-16 metres. The trunk is covered with persistent bases of petioles and the leaves are 3-4.5 metres long, greyish green in colour and have short spines at the base. The tree is deciduous and the male flowers are white, while the female flowers are greenish. Fruit is orange-yellow and an oblong-ellipsoid berry (Anon., 1988).

The origin of the species has been stated to be India-Madagascar (Ranjit Singh, 1969) or India (Pareek and Sharma, 1993) and the tree is widely grown in all tropical and sub-tropical parts of India. About 29 million palms have been reported to exist in this country (Anon., 1988).

The tree grows in tropical and sub-tropical climate but thrives better with higher productivity in sub-tropical atmosphere having rainless and dry period during development and ripening of the fruits. Soil has not been an important factor but moist alluvial soil which is not too heavy or clayee is considered to be more suitable (Anon., 1988).

Propagation of the palm takes place by seeds which germinate easily. The palm is mostly grown as stray trees but are sometimes planted along avenues or on the bank of the tanks and in such cases, transplanting of the seedlings is more commonly done.

The study of pollen grains of *P. sylvestris* in respect of germination and storage has been done by Saini and Dube (1992) and Bhattacharya *et al.* (1993) analyzed the phenolic compounds and iso-enzymes in the pollen grains. The trees produce fruits after 6 or 7 years and in West Bengal, the ripe fruits are available in summer to rainy-season (Mazumdar, 1984). In south India, the fruits are also available upto the month of October. Some diseases, *viz.*, leaf-splitting, pin-head spot, false smut, sooty mould and grey blight and a number of insect pests have been reported to attack the stem and leaves of the palm (Anon., 1988).

The fruits have been reported to contain 59.2 per cent moisture, 1.2 per cent protein, 0.4 per cent fat, 33.8 per cent carbohydrate, 3.7 per cent fibre, 1.7 per cent calcium and 0.38 per cent phosphorus (Anon., 1988). Mandal and Mazumdar (1997) had analyzed fruits harvesting from the southern plain land of West Bengal. They stated average weight per fruit as 2.39 g, specific gravity as 1.09, length as 2.47 cm, dry matter of the pulp as 0.17 per cent, titratable acidity (equivalence of citric acid) as 0.17 per cent and the vitamin C content as 20 mg per 100 g. Appreciably high amount of pectin as calcium pectate has also been reported by Majumder and Mazumdar (2001) in the pulp of the fruits.

Kapoor *et al.* (1990) determined the fatty acid composition of the seeds and stated that the major fatty acids were palmitic, oleic and linoleic. The seed oil composition of *P. sylvestris* has also been reported by Bhakare *et al.* (1993).

The sap is by far, the most important commercial product, tapped from the trunk of the palm. For tapping, the lower leaves along one side of the trunk are cut off and the inner soft layer is exposed on removing the bark. The tapping is done after a few days by making a V-shaped cut on the exposed surface and scooping out a triangular area inside that V-shaped cut. The sap exudes from the scooped surface is collected in an earthen vessel through a bamboo stick in the evening and is taken out in the early morning of the next day. A shallow cut is given in the next evening and the sap is collected. After giving rest for some days, tapping is again done and is continued over the season. The cuts are usually given at alternate sides of the trunk in a zig-zag manner. Sometimes fruit branches are cut off for optimum flow of sap. The yield of sap per palm varies from 70 to 310 kg in season (Anon., 1988).

The unfermented sap which is known as *nira* is an excellent drink in winter. However, it ferments easily within hours but fermentation can be arrested by some chemicals. The analysis of the *nira* showed a specific gravity of 1.07, *p*H of 6.75, protein of 0.37 per cent, total sugars of 11.01 per cent, reducing sugars of 0.97 per cent, mineral matter of 0.54 per cent and phosphorus of 0.16 per cent. It is also a fairly rich source of vitamins B and C.

The unfermented sap is used for making *gur*, making it alkaline by liming. To make the *gur*, the fresh sap is boiled and the scum is removed frequently. When the desired consistency is attained, it is cooled. Excellent quality of *gur* (semi-solid form) and *nabaat* (solid form) are prepared in West Bengal which are sent to other parts of India and abroad. In the central and western parts of West Bengal, more than 15 kg of *gur* is obtained from the sap, tapped from one tree (Mazumdar, 1984).

17

Phyllanthus acidus Skeels (*Averrhoa acida* Linn.) (*Cicca disticha* Linn.)

Common Names (English) : Star gooseberry, Otaheite gooseberry, Cicca etc.

Indian Names : Lavali, Noar, Narkuli, Hariphal, Cherambola, Kirnelli, Nelli, Holpholi, Rayawali, Arunelli, Harpharuri, Chalmeri etc.

Botanical Family : Euphorbiaceae.

Star gooseberry is planted in ornamental garden for its attractive foliage but the fruits are also edible. Being very sour, the fruits are not taken in fresh form but good quality pickle is made from them. A tangy jelly is also prepared by addition of high amount of sugar.

The leaves of the tree are small, arranged in two rows along small branches and some trees are deciduous (Hayes, 1970). The fruits are small, roundish but deeply lobed and the colour is light green to yellowish green.

Nativity of the tree is India where it grows as wild mostly in south and east India but partly in Bihar and eastern U.P. also.

Propagation takes place from seeds. But air-layering is also done when planted in garden. Budding and grafting are also successful.

The tree produces two crops, one in summer and the next in rainy-season. In south India however, fruits are available over a long period and almost throughout the year. Hayes (1970) stated that at Kodur in south India, some fruits are found throughout the year and in larger number in January.

Mandal and Mazumdar (1997) had physico-chemically analyzed *P. acidus* fruits, harvested in fully mature condition from trees grown as wild in the southern part of West Bengal. The average results showed fruit weight as 5.02 g, specific gravity as 1.71, stalk-stylar length as 1.5 cm, titratable acidity (equivalence of citric acid) as 2.83 per cent, vitamin   C   as 6 mg per 100 g and the total, reducing and non-reducing sugars ranging from 0.6 to 2.5 per cent. Pectin content in the fruits has also been reported by Majumder and Mazumdar (2001 b).

18

Phyllanthus emblica Linn. (*Emblica officinalis* Gaertn.)

Common Names (English) : Indian gooseberry, Myrobalan, Myrobalan plum, Emblica.

Indian Names   : Amlika, Amla, Aonla, Amlaki, Amlika, Amalka, Ambala, Amali, Usirikey, Nelli, Awala, Amalagam, Awusadanelli etc.

Botanical Family   : Euphorbiaceae.

Aonla had been given much importance in ancient India. Among plentiful benefits rendered by this tree to the human beings, the following are mentionable:

(i) Ancient Indian ayurveds had discovered a number of therapeutic properties of the fruits and other parts of this tree among whom, the name of the sage, Chyavan comes first. Many of the medicinal properties discovered those days are perhaps lost now. Nevertheless, the role of this fruit in the treatment of scurvy, diabetes, cough and cold, chronic dysentery, diarrhoea, fever, dyspepsia, jaundice, bronchitis, gum disorder etc. is well-known.

(ii) Among all human consumable fruits, this fruit ranks next to Barbados and West Indian cherry in respect of vitamin C, which is an essential nutritive component.

(iii) A number of delicious products are made from *aonla* among which the *morabba* is of great appeal. Besides this, candy, pickle, dried chips, spray-dried powder, tablets, sauce, jelly etc. are also mentionable.

(iv) Hair growth stimulating property of *aonla* fruits is well-known.

(v) Fruits are known to have contraceptive and anti-fertility properties and are used by many tribal people for the said purpose (Mazumdar, 1989).

(vi) The tree is very hardy and is planted to utilize wasteland.

(vii) It serves as a famine insurance tree by providing fruits, firewood, leaf fodder etc. (Pareek and Sharma, 1993).

The trees of *P. emblica* are evergreen in most parts of India and reach a height of even upto 20 metres in fertile soil. The trunk is like guava with smooth bark which de-barks in strips. The leaves are small and in cluster in small branches, for which it appears to be a pinnately compound leaf. The tree is monoecious. The flowers are borne in compact clusters in leaf-axils and are inconspicuous. Male flowers are light in colour and the female flowers are green which later turn brown. Pathak and Pathak (1993) have provided botanical description of various parts of the tree including the flowers. Karale *et al.* (1991) studied flowering and sex-expression of *P. emblica* in Rahuli, Maharashtra. They observed that sex expression of the branchlets showed

59.1 per cent purely male and 35.4 per cent both male and female flowers while 5.45 per cent bore no flowers. The unusually high sex ratio of male to female flowers as 26 was derived by them. Fruits are smooth, hard, greenish to pale yellow in colour and are divided into 6 segments or lobes by means of linear grooves. The stone is hexagonal having 6 small seeds.

Tropical Asia has been stated to be the origin of the species (Sham Singh *et al.*, 1967; Ranjit Singh, 1969). The trees are grown all over India and upto 1200 metres in the Himalayas and 1500 metres in the Nilgiri hills. However, eastern U.P. and especially the districts of Varanasi, Pratapgarh, Sultanpur, Rae Bareili etc. have the fame where the fruit crop covers relatively large area both as wild and in cultivation with production of best quality fruits.

The tree is grown both in tropical and sub-tropical areas but the latter with distinct summer and winter are desirable for higher productiveness. The soil is not an important factor and the trees are to some extent tolerant to salinity and alkalinity of the soil.

Propagation of *P. emblica* is mostly done from seeds under Indian condition. Seeds do not have much difficulty in germination. Nevertheless, improvement in germination as well as seedling growth by treating the seeds with GA_3, auxins, thiourea or other chemicals have been put forward by a number of workers (Dhankar and Santosh Kumar, 1996; Dhankar and Singh, 1996; Wagh *et al.*, 1998; Murugesh *et al.*, 1998; Gholap *et al.*, 2000 b). Germination improvement by treating the seeds with biofertilizer has also been reported (Seema Bhaduria *et al.*, 2000).

Root formation in stem cuttings by the use of auxins and ethrel has been stated by Sen and Bandopadhyay (1990). A number of workers have also compared different methods and time of budding and suggested the method and the date which proved to be most conducive (Pathak *et al.*, 1991; Keskar *et al.*, 1991; Singh, G. K. and Parmar, 1998).

Micro-propagation by the use of MS medium fortified with auxins, BA, 2, 4-D and kinetin etc. has been tried in *P. emblica* by some workers with success (Sehgal and Khurana, 1985; Kant and Bhanu Verma, 1997; Parul Gupta *et al.*, 1997; Bhanu Verma and Kant, 1999).

Nutritional studies in *P. emblica* have been done to a limited extent and mention may be made to the works by Sarkar *et al.* (1985) who provided results by applying N, P and K fertilizers. Proper training of the trees at early stage is essential and modified central leader system is by far, the best method. Fruit thinning is another important operation as branches may break, if there is heavy load of fruits. The trees flower in spring and the fruits are harvested in winter. In some parts of south India, fruits are available almost throughout the year. Singh, J.N. *et al.* (1989) presented physical and chemical indices of the fruits to determine maturity for harvest. The yield of a tree may be 250 kg of fruits and be higher in grafted fruits.

A number of *P. emblica* varieties have been named and are found to be grown in different parts of India but all of these could not perhaps be recognized as varieties. However, Banarsi is by far, the most famous variety. Sham Singh *et al.* (1967) mentioned 4 varieties, *viz.*, Banarsi, Green-tinged, Red-tinged and White-streaked. Under eastern U.P. condition, the varieties, Banarsi, Chakaiya, Hathijhool and some local types are grown.

Mazumdar (1986) reported analytical results for 4 varieties or types grown in West Bengal, *viz.*, Local Banarasi, Chakla (syn. Chakaiya) and two more types. Singh, I.S. and Pathak (1987) evaluated 5 varieties and mentioned suitability of the varieties for making jam, candy, drying and other products. The yield and quality of 3 varieties in Haryana have also been studied by Sharma *et al.* (1986). The description of the varieties, *viz.*, Banarasi,

Francis (Hathi Jhool) and Chakaiya has been provided by Pathak and Pathak (1993). Sanker *et al.* (1999) evaluated 7 types of *P. emblica* in Tamil Nadu.

Analysis of the fruits of *P. emblica* has been done by a number of investigators. Singh, R.R. *et al.* (1984) reported the results of fruit weight, seed weight, fruit diameter, TSS, acidity and the vitamin C done in the varieties, Banarsi and Desi. Mazumdar (1986) evaluated 4 varieties grown in West Bengal in regard to weight, size, shape, colour and texture of the fruits as well as TSS, titratable acidity and vitamin C content of the pulp. The study of the relative level of different amino acids had been done by Barthakur and Arnold (1991).

According to Gopalan *et al.* (1993), the fruits contain 81.8 per cent moisture, 0.5 per cent protein, 0.1 per cent fat, 0.5 per cent minerals, 3.4 per cent fibre, 13.7 per cent carbohydrate, 0.05 per cent calcium, 0.02 per cent phosphorus and 0.001 per cent iron. Among the vitamins, they have stated that the fruits contained 600, 0.03, 0.01 and 0.2 mg of vitamin C, thiamine, riboflavin and niacin and 9 mcg of carotene per 100 g. Mandal and Mazumdar (1995) reported the results of various physico-chemical analyses done in the fruits of a type grown in the southern part of West Bengal.

The vitamin C content in different layers of fruits had been tested by Guha and Mazumdar (1996) in West Bengal. They estimated that the vitamin C was 483.33, 512.30 and 544.70 mg per 100 g respectively in the inner, middle and outer layers of the fruits, indicating the fact that it gradually increases towards the outer side. Supe *et al.* (1998 b) reported physico-chemical characters in fruits of 5 varieties in Maharashtra. Physico-chemical characters of the fruits during development have been evaluated by Balasubramanyan and Bangarusami (1998) also and among other constituents, the vitamin C was observed to make increase steadily upto maturity.

In some parts of India, it is a common practice to store the fruits of *P. emblica* in brine water. Mazumdar (1986) studied that by steeping the fruits in aqueous solutions of common salt of 0.1, 1, 1.5, 2, 2.5 and 3 per cent concentrations for 7 days with change of solutions every alternate day, there had been a loss of vitamin C as compared to the untreated fruits and the loss had been greater with increasing strength of the solutions. The storage behaviour of the fruits by pre-harvest sprays with solutions of different chemicals had also been tested by Yadav and Singh (1999).

19

Physalis peruviana Linn.
(*Physalis edulis* Sims.)

Common Names (English) : Cape-gooseberry, Golden berry.

Indian Names : Tan kasi, Tepari, Tipari, Makowi, Rasbhori, Photi, Siruthakkali, Busarakaya, Buddabusara, Bondulla, etc.

Botanical Family : Solanaceae.

Cape-gooseberry deserves particular attention due to the fact that it is one of the few quick-growing fruit crops of India and is grown like an annual crop. Besides this, the fruits are liked by all consumers due to a good blend of sweetness and sourness and in having good flavour. Above all, excellent jam is prepared from the berries (Mazumdar, 1979 a). In some parts of the world, the fruit is used in cooking of meat as well. The fruits though soft are not damaged by birds in the field as they are enclosed by accrescent sepals. Satisfactory production of the crop is also obtained without much care.

Cape-gooseberry is a herbaceaous perennial plant attaining a height upto 80 cm. Leaves of the plants are hairy and each fruit is enclosed within persistent sepals which fuse together so as to cover fruit completely and freely. The seedlings flower in the second year.

The origin of the species is Peru. In India, it is grown all over and upto an elevation of 1800 metres in south India. The crop is commercially cultivated in north India, West Bengal and eastern India as an annual crop but as a perennial crop maintaining 2-4 years in south India, especially in the hills.

The cultivation of cape-gooseberry is done satisfactorily both in the tropical and sub-tropical situations. But the plants are much sensitive to frost and hence, the cropping season in north India is adjusted so as to avoid frost which is mainly done by early transplanting. Rain is helpful during growth but dry weather is necessary during maturation of the fruits. Soil is not considered to be an important factor for the crop but loose soil with good amount of organic matter is desirable.

Propagation of the plants is done by seeds everywhere. But vegetative propagation has been suggested where it is grown as a perennial crop (Hayes, 1970). Vegetative propagation by layering is very successful. Santana and Angarita (1997) reported micro-propagation with production of highest number of plants by culturing in the MS medium supplemented with 1 ppm of BA and GA_3 and 0.5 ppm of 2, 4-D under light.

Commercial cultivation of the crop under Indian condition has been described by Mazumdar (1979 b). Physiologically, the plant is a quantitative short day plant (Heinze and Midasch, 1991). The effect of sowing time and season of cropping on the yield of berries and their chemical composition has been studied by Poi (1989). Prasad *et al.* (1985) reported yield performance of the crop by application of N, P and K fertilizers. Poi (1989) evaluated yield and chemical composition of the berries by application of 3 levels of N, P and K fertilizers, either singly or in all possible combinations.

The effect of different types of manure, fertilizer and soil-amendments in cape-gooseberry had been studied by Wolff (1991) and among other findings, no effect on fruit quality by application of the amendments was noted by them.

The yield and composition of the fruits by application of different plant growth regulators has been studied by Poi (1989). The same workers also established the level of irrigation for highest yield and quality of the berries.

Fischer and Martinez (1998) studied on the quality and maturity of the fruits in relation to their change of colour from green to reddish orange and opined that skin colour could be used in the field as an indicator of quality and maturity. Scheer et al. (1999) had also studied on correlation between skin colour and quality of the fruits. They had put forward that evaluation of the crop during harvest and also marketing could be reliably conducted by non-destructive colour determination.

As regards composition of the fruits, Gopalan et al. (1993) has mentioned that the fruits contained 82.9 per cent moisture, 1.8 per cent protein, 0.2 per cent fat, 0.8 per cent minerals, 3.2 per cent fibre, 11.1 per cent carbohydrate, 0.01 per cent calcium and phosphorus and negligible iron. The fruits are known to be rich sources of bioflavonoids as well (Hayes, 1970).

The changes in different physico-chemical characteristics of the fruits during development and maturation had been worked out by Poi (1989) under West Bengal condition and similar study had been done later by Sarangi et al. (1992) in the same experimental situation. On studying developmental changes of carbohydrates in the fruits, Fischer and Ludders (1997) opined that starch was most abundant during all stages of their development. The said workers [Fischer *et al.* (1997)] had later reported the changes that took place in regard weight, size, soluble solids, acidity and the beta-carotene content in the fruits during development. The activity of different carbohydrate enzymes in the fruits during development had been reported by Trinchero *et al.* (1999). It has been pointed out by them that ripening of the fruits was associated with a marked climacteric rise in carbon di-oxide and ethylene production.

The effect of different concentrations of IBA and GA_3 on changes in the pectic polysaccharides besides activities of the enzymes, *i.e.*, pectin methyl esterase (PME) and polygalacturonase (PG) were assayed by Majumder and Mazumdar (2001) in the fruits of cape-gooseberry during development and ripening. It was observed that the water-soluble pectic fractions accumulated more in the auxin-treated fruits than the control and GA_3 had raised the level of the acid-soluble and alkali-soluble fractions. Besides, auxin enhanced polygalacturonase activity while GA_3 highly reduced the activity. The significance of such results in relation to pectic solubilization has been discussed by the workers.

The said authors had later (2002) put forward that while the water-soluble and the oxalate-soluble pectic substances increased during ripening, the acid-soluble and alkali-soluble fractions decreased. Simultaneously with degradation of high molecular weight pectin, there was a 5-6 fold increase in PG activity while PME activity was not clearly related to ripening. The increased level of PG activity was highly correlated with ethylene evolution occurred prior to PG synthesis in the fruits. The possible mechanism of pectic changes in the berries has been elaborated (Majumder, 1999).

The yield of cape-gooseberry greatly varies and may be upto 25 tonnes per hectare. Any serious pest is hardly met with in the crop, although mosaic virus is sometimes noted,

Little leaf disease frequently becomes a problem in West Bengal and Poi *et al.* (1987) reported suppression of the symptoms with production of normal healthy leaves by 4 sprays with tetracycline solutions.

20

Psidium cattleianum Sabine (*Psidium littorale* Raddi)

Common Names (English) : Cattley guava, Strawberry guava, Purple guava etc.

Indian Names : Lal amrud, Lal anjir, Beguni peyara etc.

Botanical Family : Myrtaceae.

Cattley guava is an ornamental tree and is a relative of guava. Although the fruits which are small and are not comparable to guava in quality, they are rich sources of pectin of high jelly-grade and hence, serve as excellent materials to make jelly of high quality.

Two botanical varieties or sub-species of *P. cattleianum* are recognized, which are (i) *P. cattleianum* var. *lucidum*, fruits of which are yellowish and roundish and (ii) *P. cattleianum* var. *acre*, having elongated fruits while the fruits of *P. cattleianum* are purplish red.

The tree of *P. cattleianum* is shrubby in habit and the height goes upto 7 or 8 metres in the tropics. The tree however, makes slow growth. Leaves are deep green, leathery, obovate to elliptic in shape and are 8-10 cm long. The fruits of the yellow-fruited *lucidum* are sweeter than the purplish fruits.

In India, the tree is found to grow largely in south Indian plains and upto 1600 metres in the Nilgiris. The tree is also found to grow in eastern India.

Normand (1994) made a botanical and horticultural review of strawberry guava and mentioned that in Réunion island, it grows on the windward side of the island from sea-level to 1300 metre alttitude. He added that it would be a suitable fruit crop there for the wet highlands.

Propagation is done mainly from seeds. Voltolini and Fachinello (1997) reported success of hardwood cuttings by suitable application of IBA and especially when stock plants were grown under 70 per cent shade.

In south India, the fruits are available in the months of July-August. Mukai *et al.* (1989) from Japan reported that harvesting time for both *P. cattleianum* and the var. *lucidum* was earlier at higher temperature. They mentioned that fruit weight was higher at high temperature in the yellow type but higher at lower temperature in the red type. They opined that fruit maturation of the red type was affected by temperature and fruit quality was better when the plants were grown at 20°-25°C. The fruit characteristics in respect of TSS, acidity, anthocyanin, carotenoids and sugars have been studied by them.

According to Gopalan *et al.* (1993), *P. cattleianum* fruits contain moisture 85.3 per cent, protein 0.1 per cent, fat 0.2 per cent, minerals 0.6 per cent, fibre 4.8 per cent, carbohydrate 9.0 per cent, calcium 0.05 per cent, phosphorus 0.02 per cent and iron 0.001 per cent. Vernin *et al.* (1998) reported a number of volatile compounds in the fruits of *P. cattleianum* and *P. cattleianum* var. *lucidum* grown in Réunion island.

Galho *et al.* (2000) in Brazil studied fruit development of a cultivar of *P. cattleianum* and found that development was completed in four months. They added that dry matter accumulation increased until the end of fruit development and the greatest rate of dry matter accumulation had occurred at 67 days after anthesis. However, the length, diameter, volume and fresh matter of the fruits showed a biphasic pattern described by a double sigmoid curve.

21

Spondias cytherea Sonner.
(*Spondias dulcis* Sol. ex Forst. f.)
and *Spondias pinnata* (L. f.) Kurz.
(*Spondias mangifera* Willd.)
(*Spondias acuminata* Roxb.)
(*Mangifera pinnata* L. f.)

Common Names (English) : Hog plum, Otaheite plum,
Golden apple, Ambarella,
Otaheite apple (for S. *cytherea*
Sonn.) and Indian hog plum
(for S. *pinnata* Kurz).

Indian Names   : Amrataka, Ambula, Amra,
Bilati amra, Mumbulichi,
Mampuli, Bahamb, Jangliam,
Ranamba,   Marima,
Ampalam,   Ambalamu,
Amate,   Konda-mamidi,
Ambarela, Pundi etc.

Botanical Family   : Anacardiaceae.

Fruits of these species are not eaten as raw for high acerbity but a number of products can be made from them such as pickle, sour soup, *chutney* etc. Squash and jam can also be prepared from the fruits, especially with those of *S. cytherea* by suitably blending with other fruits. Fruits of this species are somewhat sweet, less sour and hence, these are also relished with addition of salt, chilli powder and other spices. The trees of both the species are planted along avenues as well.

The origin of the species, *S. cythera* has been stated to be Polynesia while that of *S. pinnata* in tropical Asia (Ranjit Singh, 1969). Nativity for the latter species in India has also been mentioned (Pareek and Sharma, 1993). In India, the trees of both the species are grown as wild or are sometimes planted in household compounds in West Bengal, Assam, all over the south India and parts of Maharashtra.

The trees of *S. pinnata* are highly spreading and may reach height similar to that of mango trees grown from seeds, while those of the other species are not so tall and spready. They are deciduous in nature. The flowers are polygamous with 8-10 stamens and 4-5 styles. Fruits are oval and oblong in shape and measure 6-8 cm in length for *S. pinnata* and larger for those of the other species. On full ripening, fruits of both the species develop rich or pale yellow colour of the peels.

The trees of both the species are more suitable to tropical climate. They also grow in the moist sub-tropics but growth and productivity are usually lesser. In low rainfall area, fruit drop is high. Soil is though not an important factor, the trees cannot withstand poor drainage.

According to Gopalan *et al.* (1993), the fruits of *S. pinnata* (*S. mangifera*) contain 90.3 per cent moisture, 0.7 per cent protein, 3.0 per cent fat, 0.5 per cent minerals, 1.0 per cent fibre, 4.5 per cent carbohydrate, 0.036 per cent calcium, 0.011 per cent phosphorus and about 0.004 per cent iron. Nahar *et al.* (1990) observed that fruits grown in

Bangladesh contained substantial amounts of dry matter, ash, lignin, starch and dietary fibres and that glucose and fructose were the main components. The physico-chemical analyses done by Mandal and Mazumdar (1995) with fruits of *S. cytherea* harvested in the month of August from trees grown in the central part of West Bengal showed average weight per fruit as 78 g, specific gravity as 1.04, stalk-stylar length as 6.2 cm, total sugars as 2.58 per cent, reducing sugars as 1.67 per cent, non-reducing sugars as 0.86 per cent, titratable acidity (citric acid equivalence) as 0.85 per cent and vitamin C as 50 mg per 100 g. Appreciable amount of pectin as calcium pectate in the fruits of *S. cytherea* grown in West Bengal condition has also been estimated by Majumder and Mazumdar (2001 b).

On analyses of the fruits of *S. pinnata* done at three stages of development, *i.e.*, at the end of April, early part of May and middle of June under the conditions of West Bengal, Bhattacharyya *et al.* (1992) reported gradual increase of the total soluble solids, total reducing and non-reducing sugars as well as vitamin C in the fruits with progress of their development. The acidity made an increase initially but it decreased later though not significantly. With fruits of *S. cytherea* sampled from trees grown in the central part of West Bengal, Mandal and Mazumdar (1998) determined average weight, specific gravity, length, diameter and the amount of total, reducing and non-reducing sugars, titratable acidity and the vitamin C at four equally spaced stages of their development such that the last sampling date coincided with harvest maturity under local conditions. It was observed that while acidity dropped down, other constituents made steady increases with development though at different degrees. The increase had occurred at greater extent for dry matter, sugars and the sugar-acid ratio at the advanced maturity condition of the fruits. The changes in other components of the fruits had also been studied by Sinha Roy (1996) and Mandal (1997).

22

Syzygium cumini (Linn.) Skeels
(*Myrtus cumini* Linn.)
(*Eugenia jambolana* Lamk.)

Common Names (English)	: Black plum, Java plum, Indian blackberry, Jambolan etc.
Indian Names	: Jambu, Jamun, Jaman, Jam, Kalo jam, Jambul, Phalani, Phalinda, Ra jamun Jamukoli, Jamdudo, Jam buvu, Naval, Neredu, Neralu, Mahadan etc.
Botanical Family	: Myrtaceae.

Organized culture of black plum in the orchard is although rarely met with under Indian conditions, it is planted in home lands, public compounds, along avenues, in forests, as wind-break and also grow as stray trees in large number. Hence, production of this fruit is considerably high in India. In the recent times however, plantations of black plum are also coming up in some parts of India.

Ripe fruits of this species are highly relished to both rich and poor people in this country and accordingly, they have a great demand in season of availability. A number of products are also made from this fruit such as jelly, jam, wine, vinegar, pickle etc. but squash or sherbet is by far, most widely used.

However, black plum is highly valued in India for a number of medicinal properties in its fruits, seeds and leaves and are recommended to control diabetes, dysentery, diarrhoea, oedema, ringworm, fever etc. (Hayes, 1970; Chandra, 1985; Al-Zaid et al., 1991; Chaudhuri et al., 1990; Chowdhury, 1992). Apart from anti-diabetic preparations, the seeds are also known to be very good concentrate feed to the cattle as they are rich in protein, carbohydrate and calcium (Hayes, 1970).

The trees are robust, evergreen, spready and attain a height of 10 metres or more. Flowers are borne in leaf-axils and are bisexual and light yellow in colour. Misra and Bajpai (1975) described floral biology and anthesis of *S. cumini*. The fruits are oval or obovate and may be upto 6 cm long borne by some trees under specific agro-climatic conditions. On ripening, the peel becomes blackish purple and the pulp may be white, pink, red or purple, depending on types.

The origin of *S. cumini* has been stated to be India, south-east Asia, Malaya, Burma, Sri Lanka etc. (Ranjit Singh, 1969; Hayes, 1970; Pareek and Sharma, 1993). In India, the trees grow successfully in all tropical and sub-tropical parts and also in the Nilgiris and the Himalayan region upto about 1200 metres.

The trees are resistant to drought. Dry atmosphere during flowering and fruiting enhances productivity of the fruits with higher sugar accumulation. Soil is not important but poor drainage greatly damages the tree. Gurbachan Singh et al. (1998) in Haryana studied the response of *S. cumini* trees to soil pH of 8.1, 8.4, 8.7, 9.4 and 10.0. Among other observations, it was noted that after 90 days of planting, the trees showed 60 per cent survival at pH 9.4.

Seed propagation is very common in India but the seeds have low viability period. Polyembryony is very often noticed in the seeds. Mazumdar (1990) observed that out of 200 seeds sown, 31.9 per cent were polyembryonic which protruded upto 4 seedlings per seed. Among other workers, polyembryony has also been reported by Kader *et al.* (2000). As regards sowing depth, Sultan Singh and Singhrot (1984) stated that maximum germination of 78 per cent as well as best seedling growth was obtained when sowing was done at 5 cm depth. The effect of the date of sowing on germination has also been stated by the same workers (Sultan Singh and Singhrot, 1985).

S. cumini is considered to be a difficult-to-root species. Sircar (1986) also stated that neither forced cuttings nor normal cuttings had produced adventitious roots. Air-layering, if done in rainy season by the use of auxin is found to produce roots in many cases. However, success by adopting different methods of budding and grafting have been on record (Rema and Krishnamoorthy, 1994; Bhagat *et al.*, 1999; Chovatia and Singh, 2000 a). Yadav and Singh (1999) reported to have obtained multiple shoots from nodal and shoot tip segments of 10 to 15 day-old seedlings of *S. cumini* on modified MS medium supplemented with suitable concentrations of BA, singly or in combination with NAA, IAA or IBA. The regenerated plantlets were acclimatized and successfully transferred to soil. Micro-propagation of *S. cumini* has also been reported by Roy *et al.* (1996) using MS medium supplemented with kinetin.

The trees are usually not grown by application of any agro-input but organic manure and especially potassium fertilizer greatly improves yield and quality of the fruits. Balakrishnan *et al.* (2000) in Madurai studied the effects of iron, zinc and magnesium deficiencies on pigment composition, nutrient content and photosynthetic activity in field grown trees of *S. cumini.*

In attempts to determine the possible causes of low seed set in *S. cumini,* Arathi *et al.* (1996) concluded from their study that post-fertilization factors might be involved as pollen grain limitation and lack of fertilization did not completely explain the single seededness of the fruits. Harvesting of the fruits of *S. cumini* in most parts of India is continued for a month and is done in the months from June to August. In south India, two crops are also obtained.

Any standard variety of *S. cumini* is not known. But *Ra Jamun* can be named which is a common type rather than a variety grown in north India. The fruits of this variety are of fairly large in size, having deep purple peel, pinkish pulp, small seed and claim to be more sweet in taste than other types.

According to Gopalan *et al.* (1993) *S. cumini* fruits contain 83.7 per cent moisture, 0.7 per cent protein, 0.3 per cent fat, 0.4 per cent minerals, 0.9 per cent fibre, 14.0 per cent carbohydrate, 0.015 per cent calcium, 0.015 per cent phosphorus and 0.004 per cent iron. A number of sugars and galacturonic acid had been extracted from the pulp and seeds of *S. cumini* by Gomes *et al.* (1986). Mandal and Mazumdar (1997) had analyzed *S. cumini* fruits grown in the southern part of West Bengal and presented data for different physical and chemical components. Appreciable amount of pectin as calcium pectate has also been observed by Majumder and Mazumdar (2001 b) in *S. cumini* fruits grown in the southern part of West Bengal.

The changes in the photosynthetic pigments and sugars in the fruits at four stages of development have been determined by Rao and Charyuhi (1989) who observed various increases and decreases of the components. Among other investigators who studied changing pattern of different components in the fruits during their development, mention may be made to the works by Venkitakrishnan *et al.* (1997) and Garande *et al.* (1998 a, b). The activity of seed extracts of S. *cumini* and the fatty acids present in the seed oil have respectively been put forward by Chakraborty *et al.* (1986) and Daulatabad *et al.* (1988).

23

Syzygium jambos (Linn.) Alston (*Eugenia jambos* Linn.)

Common Name (English) : Rose apple.

Indian Names : Golap jam, Gulab jamun, Paninirchampa etc.

Botanical Family : Myrtaceae.

Rose apple is an excellent fruit and appeals everybody for its nice rosy fragrance, good and spongy texture which is neither very soft nor very hard and very low acerbity. The fruits are also used for canning, jellying or for candying. The new leaves, greenish white flowers with long stamens and the spreading habit of the trees are also attractive for which they are selected for planting in many ornamental gardens and along avenues. Medicinal value of the fruits is also known (Campelo, 1988).

The origin of the species is India or south-east Asia. The tree is grown as wild or are also planted in West Bengal, Assam and Orissa. It is however, grown abundantly in south India and upto an elevation of 1350 metres (Hayes, 1970).

The trees are spreading and reach a height of 8 metres or more. The leaves are thick and shiny, oblong-lanceolate in shape, long and may be 18-20 cm. The fruits are almost spherical, 3-5 cm in diameter but some trees produce very large fruits. Polyploidy of the species has also been indicated (Hayes, 1970). The seed is large, brown, numbering 1-3 and are enclosed in large cavity of each fruit.

S. jambos is adaptable to both dry and humid climate, but grows better in tropical climate. Yamada *et al.* (1996) in Japan evaluated the heat tolerance of the species by determining chlorophyll fluorescence and observed that the trees were sensitive.

Propagation of *S. jambos* is normally done by seeds. The seeds are polyembryomic and two or more plants emerge out from a seed (Sham Singh *et al.*, 1967; Hayes, 1970).

However, Sharma *et al.* (1989) reported success with cuttings and best rooting (96.8 per cent) was observed by them on treating the cuttings with 2500 ppm of NAA combined with 1000 ppm of ethephon. Rooting of cuttings by the use of growth regulators as also seasonal influence has been studied by Hore and Sen (1994). Air-layering of *S. jambos* has been experimented by Martins and Antunes (2000) in Brazil on application of various concentrations of IBA. Beneficial effect of rooting by application of higher concentrations of the auxin has been indicated by these workers. Litz (1984) studied the *in vitro* responses of adventitious embryos, culturing on modified MS medium.

As regards nutrient uptake by the trees, the study made by De *et al.* (1996) had aimed at determining the nitrogen, phosphorus and potash content of the leaves explanted from trees growing in the central part of West Bengal. Analyses were done by them at monthly interval from December to May and it was observed that regardless of the elements, there had been an initial decline of the elements which tended to rise later with advancement of summer season. In Germany, the combined effect of water and nitrogen on growth and gas exchange characteristics of *S. jambos* had been undertaken by El-Siddig *et al.* (1998). Among other observations, the investigators had noted that addition of nitrogen had a compensating effect on the reduction of dry matter production in presence of water stress.

With experiments done in breeding biology, Chantaranothai and Parnell (1994) in Thailand noted that the species was self-compatible and self-pollination was of common occurrence. According to them, seed set via apomixis, autogamy and geitonogamy appeared to be enhanced by pollination. Experiments done by Mukhopadhya (1992) on parthenocarpy of *S. jambos* revealed that among other auxins, NAA at 1000 ppm applied to the stigma and style of flowers gave 100 per cent success.

Fruit analytical report of *S. jambos* as presented by Gopalan *et al.* (1993) is as follows. Moisture 89.1 per cent, protein 0.7 per cent, fat 0.2 per cent, minerals 0.3 per cent, fibre 1.2 per cent, carbohydrate 8.5 per cent, calcium 0.01 per cent, phosphorus 0.030 per cent and iron 0.005 per cent.

Physico-chemical analyses of the *S. jambos* fruits grown in the central part of West Bengal have been done by a number of workers. Mazumdar (1979 b) reported average weight per fruit as 11.38 g, specific gravity as 0.16, total titratable acidity (equivalence of citric acid) as 0.10 per cent and the total, reducing and non-reducing sugars respectively as 4.58, 4.11 and 0.44 per cent. Subsequently, Sinha Roy (1996), Mandal (1997) and Mandal and Mazumdar (1997) put forward the results obtained on analyses of the fruits done in respect of average weight, specific gravity and length per fruit as well as the percentage of titratable acidity, pectic compounds, total, reducing and non-reducing sugars, vitamin C and the dry matter content of the pulp. The said workers (Mandal, 1997; Mandal and Mazumdar, 2000) later sprayed *S. jambos* trees with 1 per cent solution of KCl or $ZnSO_4$, or 50 ppm solutions of IBA or GA_3 after 12-15 days of fruit set. The treated or untreated fruits were plucked at harvest-maturity and analyzed for weight, specific gravity, length, diameter and the content of dry matter, total, reducing and non-reducing sugars, titratable acidity and the vitamin C. All the treatments were observed to have enhanced the level of the different constituents over control except that the specific gravity, non-reducing sugars and the vitamin C content, in which case, any effect of the chemicals had not been significantly pronounced. The effect of application of the said chemicals on pectic fractions in the fruits was also studied by Sinha Roy (1996).

Chattopadhyay and Ghosh (1996) sampled fruits from bud stage to harvest-maturity and analyzed them for the concentrations of nitrogen, phosphorus, potassium and iron at different stages of their development. It was observed that the nitrogen content reached the highest level at 20 days after fruit set, phosphorus and potassium were higher at maturity than during early stage of growth and iron content was highest in 30 day-old fruits. Physico-chemical analyses of the fruits at four stages of development have later been done by Mandal and Mazumdar (1998) for different components.

24

Syzygium samarangense (Blume) Merrill & Perry (*Eugenia javanica* Lamk.) (*Myrtus samarangensis* Bl.)

Common Names (English)	: Water-apple, Malay-apple, Wax-apple, Star-apple etc.
Indian Names	: Jamrul, Sada jam, Safed jamb, Pani jamut, Malay jaman etc.
Botanical Family	: Myrtaceae.

S. samarange is grown as stray trees in tropical parts of India. In West Bengal however, it is also planted in household and public compounds, where the fruit has a great demand and is regarded by many as the best fruit to quench thirst in hot weather. For the attractive appearance of the fruits and especially those that are of pink colour, the tree is sometimes planted in ornamental gardens also. Medicinal value of the fruit is also on record (Campelo, 1988).

The tree of *S. samarangense* is tropical and is found to be at its best where rainfall is widely distributed along with mild winter and summer. Yamada *et al.* (1996) in Japan studied heat tolerance of the trees of *S. samarangense* by determining chlorophyll fluorescence and found those to be very much sensitive to heat. Although soil is not considered to be an important factor, the tree is not found to grow satisfactorily in poorly drained soil in West Bengal. Hsu Yu Mei *et al.* (1999) evaluated the changes in carbon and nitrogen metabolism in response to flooding of 2 or 3 years old potted tree of *S. samarangense* grown outdoors in Taiwan. It was observed by them that in leaves, starch content markedly increased after 14 days and the total nitrogen decreased after 35 days of flooding. Besides, the TSS in the roots had significantly increased after 14 days. The data on changes occurred in the content of carbohydrate, soluble protein, amino acids and the activity of glutamine synthetase had also been presented by the said authors.

Propagation of the trees is mostly done by seeds. Sarkar *et al.* (1984) reported root formation in hardwood stem cuttings by combined use of phenolic compounds and auxins. *In vitro* responses of adventitious embryos had been studied in *S. samarangense* by Litz (1984). Immature adventitious embryos from excised ovules from immature fruits were cultured by them in MS medium. Depending on developmental stage of the embryos, proliferation of axillary buds had been observed in the media with 2-10 mg/litre of BA. Root formation was induced only on media that contained 3-10 mg of 2, 4-D litre. Embryonic callus had been found from adventitious embryos from 1-2 cm fruitlets on 1-2 mg/litre of 2, 4-D.

Wattanawikkit *et al.* (1995) in Thailand had cultured axillary buds on Woody Plant Medium (WPM) supplemented with various concentrations of BA. The growth regulator at 9 mg/litre was found to produce highest average shoot number while the medium without any regulator gave highest average shoot length after 120 days of culture. Highest root induction percentage (75 per cent) was obtained at 4 mg of IBA/litre for 60 days. Plantlets dipped in 5 mg of IBA/litre before transplanting brought about 90 per cent survival.

In study related to nutrition of the trees, De *et al.* (1996) determined the quantitative shift in the nitrogen, phosphorus and potash content in the laminar tissues at monthly interval from December to May under the conditions of West Bengal. Steady increase had been found to take place upto the month of February for phosphorus and March for the other constituents, following which there had been a sharp decline with advancement of summer.

On conducting experiments in pollination of *S. samarangense,* Chantaranothai and Parnell (1994) concluded that the species was self-compatible and self-pollination had been more common. Ants, honey bees and other insects were also noted by them in day time but of a small number only.

Analysis of the fruits of *S. samarangense* has been done by some workers. Mazumdar (1979 b) stated average weight per fruit as 42.5 g and the percentage of titratable acidity (equivalence of citric acid), total, reducing and non-reducing sugars of the pulp had been determined as 0.08, 5.02, 4.7 and 0.3 respectively. The volatile compounds in the fruits were isolated by Wong and La1 (1996), according to whom 36 and 41 constituents were identified in fruits of two cultivars. Increase in weight, specific gravity, length and diameter of fruits as well as in the percentage of dry matter, total, reducing and non-reducing sugars, titratable acidity and vitamin C of the pulp by foliar spraying the trees with 1 per cent solution of KCl or $ZnSO_4$ or 50 ppm of IBA or GA_3 after 12-15 days of fruit set had been observed by Mandal and Mazumdar (2000). Among other treatments, $ZnSO_4$ solution proved to be much responsive to heighten the constituents. The pectin content of the fruits as calcium pectate had been reported by Sinha Roy (1996) and Majumder and Mazumdar (2001 b).

Shu Zen Hong (1999) in Taiwan observed that in the pink-fruited type, position of fruits on the tree had a marked influence on fruit quality. It has been observed by them that the fruits on the lower trunk were heaviest and largest among the 9 positions compared. The upper inner fruits were though smallest but had a more intense red colour. They opined that for high TSS content, harvesting from lower inner positions should be done. Later, the same worker with others (Shu Zen Hong *et al.*, 2001) studied the effects of light, temperature and sucrose on colour, weight, diameter and soluble solids of the skin of the pink-coloured fruits by culturing skin discs. It was noted that light with 20°C and 6 per cent sucrose gave highest soluble solids concentration and anthocyanin content.

The quantitative changes in weight, specific gravity, length and diameter of the fruits along with percentage of dry matter, sugars, acidity and vitamin C content in the fruit pulp of the white variety grown in West Bengal have been determined by Mandal and Mazumdar (1998) at 4 stages of development at an interval of 11 days till harvest-maturity condition. Marked increase in the chemical components except vitamin C had been observed by them at the advanced maturity condition of the fruits.

25

Syzygium uniflora Linn.

Common Names (English) : Surinam cherry, Brazil
                         cherry, Pitanga cherry etc.
Indian Name   : (Not known).
Botanical Family   : Myrtaceae.

Surinam cherry has its value as an ornamental shrub
and also for edible fruits. New shoots produced by the shrub
are wine-coloured and the red, angular fruits are also of
attractive appearance. The fruits have pleasant aroma and
although can be eaten afresh, more commonly jelly and
drinks are made from the soft pulp of them. According to
Oguntimein and Elakovich (1991), the fruits are used in
Nigerian folk medicine while Ferro *et al.* (1988) mentioned
that leaves of the shrub are used in Paraguay for lowering
cholesterol, weight, uric acid, blood pressure etc.

The species is of Brazilian origin and in India, it grows
in the Nilgiri hills of south India at altitudes from 458 to
1678 metres (Sham Singh *et al.*, 1967). Recently, it has been
introduced in other parts of India and particularly to plant
in the decorative gardens. According to Hayes (1970), it is
also grown in the plain land of north India. The height of

the shrub may go upto 7 metres or more in fertile soil. Leaves are glossy, green to reddish in colour and are evergreen. Fruits are sub-globose, measure 2-4 cm in diameter, hang in cluster on slender stems, have 8 ribs and are with very thin skin. The colour of the fruits is deep crimson or almost black when ripe. The pulp is soft, juicy, deep red in colour and has pleasant aroma. Each fruit contains a large round seed but two seeds are also infrequently observed.

As compared to the related species, *S. uniflora* is able to withstand cold better (Sham Singh *et al.*, 1967). Soil is though not important but growth of the shrub and yield are seen to be prolific when grown in rich soil and of good drainage.

Propagation is done by seeds but cleft grafting and side grafting are also done in other countries. Bezerra *et al.* (1999) made study on budding and grafting of *S. uniflora* in Pernambuco, Brazil. The seedlings were grafted by cleft or whip method on to 6, 9 or 12 month old rootstocks or budded by patch or shield budding on to 12, 15 or 18 month old rootstocks. It was observed that cleft or whip grafting on 9 or 12 month old rootstocks resulted in a high percentage of take and shield budding gave better result than patch budding.

The report by Miller and Bazore (1945) had stated that the pulp of Surinam cherry fruits contained 22 per cent carbohydrate mostly as sugar and a large amount of acid also, giving a *p*H value of 2.7. A number of biochemical and pharmacological studies of fruits or other plant parts of the species have also been on record. Mention may be made to the study done by Schmeda *et al.* (1987) with leaf extracts which showed inhibitory effect against the enzyme, xanthine oxidase Adebajo *et al.* (1989) worked on anti-microbial activities and microbial transformation of the volatile oils of the Surinam cherry fruits. According to them, the results obtained may provide explanation for the use of the species in folk medicine against disorders of the digestive tract.

26

Trapa natans Linn. var. *bispinosa* (Roxb.) Makino (*Trapa bispinosa* Roxb.) and *Trapa natas* Linn. var. *quadrispinosa* (Roxb.) Makino (*Trapa quadrispinosa* Roxb.)

Common Name (English)	: Water-chestnut.
Indian Names	: Singhara, Paniphal, Singada, Shingori, Shengoda, Singara kottai, Kubyakam, Karimpola etc.
Botanical Family	: Onagraceae.

Water-chestnut is an aquatic nut crop of India, the kernels of which are relished afresh or boiled or dried and ground to make a number of products (Mazumdar, 1985). The kernels have many medicinal properties as well (Keosuge *et al.*, 1986; Sugiyama, 1988).

The plants are floating annuals, the rosulate leaves of which have inflated petioles. Flowers have no inner whorl of stamens, ovary is half inferior, bilocular with a single pendulous ovule in each locule. The floral biology of the plant has been studied and described by Kadono and Schneider (1986) and Susumu Arima *et al.* (1999) in Japan. The fruit is a one seeded, top-shaped drupe, the fleshy pericarp of which is deciduous and covers a large 2 or 4 horned stony endocarp (pyrene). The seed is dicotyledonous and the cotyledons are unequal in size such that one is very large while the other is minute and scale-like. The horns represent sepals.

The nut crop is cultivated in the ponds in all tropical and sub-tropical parts of India. In Japan, it is also grown as wild (Arima, 1994).

Mazumdar (1985) has described cultivation of water-chestnut under Indian conditions.

Propagation of the plant is commercially done by seeds. The fully mature nuts are placed in container with little water to germinate the seeds. The sprouted seeds are sorted out and broadcast in nursery tanks. At the beginning of monsoon, the seedlings are lifted from the nursery tanks and planted in pond, at spacing of 1-2 metres or 2-3 metres when the soil of the pond is fertile.

The seeds of water-chestnut are recalcitrant and cannot be stored long. Cozza *et al.* (1994) studied the effect of storage of the nuts at low temperature on germination and viability and they have stated that maximum germination had occurred at 15°-25°C.

Micro-propagation of water-chestnut has also been on record. Zhou *et al.* (1983) reported production of plantlets by using MS medium supplemented with suitable concentrations of IAA or GA and BA. *In vitro* germination and micro-propagation has also been reported by Anuradha Agrawal and Ram (1995) using Nitsch's basal semi-solid medium.

In West Bengal, some growers apply 30-40 kg of urea per hectare area of pond after about a month of transplanting and the same dose is applied 20 days later. The effect of nitrogen application on the growth and yield of the crop has been studied by Arima *et al.* (1989) in Japan. They have also noted that application of 35 kg of NPK fertilizer had stepped up fruit yield by about 100 per cent over the untreated control.

Lessening of water in the pond due to drought may create difficulty and in such case, it should be replenished with water from other source. Luxuriant vegetative growth of the plants as may result in highly fertile condition of the medium lowers productivity of the plants and hence, mild pruning becomes necessary in such case. Regular eradication of aquatic weeds, especially, *Hydrilla* and *Eicchornia* becomes necessary during the cropping season.

No standard variety of water-chestnut is known but nuts having husks of green, red or purple and a blending of red and green colour are recognized. Kanpuri, Jaunpuri, Desi large, Desi small etc., are the names of some types of water-chestnuts referred to by the growers in West Bengal and other parts of eastern India.

Analysis of the nuts has been done by a number of workers. Under West Bengal condition, the physico-chemical analyses of *T. natans* var. *bispinosa* in respect of weight and size of nuts and kernels and the content of dry matter, carbohydrate, sugars, acidity, vitamin C, protein, ether-extractive part and pectin of the kernels belonged to the green and purple-husked nuts have been reported by some workers (Mazumdar and Jana, 1977; Pratima Roy and Mazumdar, 1989; Ali and Mazumdar, 1990, 1991; Mandal and Mazumdar, 1995). Mandal *et al.* (1997) collected green and red-husked nuts of *T. natans* var. *bispinosa* and green-husked nuts of *T. natans* var. *quadrispinosa* from different parts of West Bengal and analyses of those proved superiority of the former type in respect of dry matter, carbohydrate and sugars although the latter had higher kernel weight.

The results on the changing pattern of different physico-chemical components at 4 stages of development till attainment of harvest-maturity have been provided by Mandal (1997) and Mandal and Mazumdar (1998) with green-husked nuts of *T. natans* var. *bispinosa* grown in the southern parts of West Bengal.

Although disease is not serious, 3 or 4 insect pests and snails damage the crop and sometimes to a great extent. Studies have been done to test the effect of Monocrotophos, Carbaryl and other insecticides in controlling *Galerucella birmanica* (Yadav and Gargav, 1988) and the effect of Monocrotophos in controlling the larvae of *Stenochironomus* species (Yadav, 1992) in Madhya Pradesh. The biology and control of the former species have also been studied by Lu *et al.* (1984) in China.

Harvesting of the nuts is usually started from the month of September and continues upto November. For the purpose of harvesting, specially made rafts are used by the growers. Observational trial done by Mazumdar (1982) in the southern part of West Bengal revealed that the yield of fresh nuts range between 2500-3800 kg per hectare area of pond but this had gone upto more than 5000 kg per hectare by applying about 50 kg of urea per hectare of pond along with eradication of weeds.

Literature Cited

ADEBAJO, A.C.; OLOKE, K.J.; ALADESANMI, A.J. (1989). Antimicrobial activities and microbial transformation of volatile oils of *Eugenia uniflora*. *Fitoterapia*, **60**(5): 451-455.

AKINWALE, T.O.; ALADESUA, O.O. (1999). Comparative study of the physico-chemical properties and the effect of different techniques on the quality of cashew juice from Brazilian and local varieties. *Nigerian Journal of Tree Crop Research*, **3**(1): 60-66.

ALAM, M.; SIDDIQUI, M.B.; HUSAIN, W. (1990). Treatment of diabetes through herbal drugs in rural India. *Fitoterapia*, **61**(3): 240-242.

ALI, S.L.; MAZUMDAR, B.C. (1990). Studies on pectin content of some fruits and plant parts. *Science and Culture*, **57**(10-11): 256.

ALI, S.L.; MAZUMDAR, B.C. (1991). Protein and fat content of water-chestnut (*Trapa bispinosa*). *Indian Agriculturist*, **35**(4): 259-260.

ALLOUPA, M.A.I.; CAMPOS, M.deA. (1990). [Callogenesis of acerola (*Malpighia glabra* L.) from leaf explants cultured *in vitro*.] *Revista Brasileira de Fruiticultura*, **21**(3): 284-287.

AL-ZAID, M.M.; HASSAN, M.A.M.; BADIR, N.; GUMAA, K.A. (1991). Evaluation of blood glucose lowering activity of three plant diet additives. *International Journal of Pharmacognosy,* **29**(2): 81-88.

AMIN, M.N.; RAZZAQUE, M.A. (1993). Regeneration of *Averrhoa carambola* plants *in vitro* from callus cultures of seedling plants. *Journal of Horticultural Science,* **68**(4): 551-556.

ANAYA HERNANDEZ, F.J.; VEGA CUEN, A. (1992). [Sexual propagation of star apple (*Chrysophyllum cainito* L.) in Morelos State, Mexico.] *Revista Chapingo,* **15**(73-74): 97-100.

ANONYMOUS (1988). The *Wealth of India-Raw Materials.* Publication and Information Directorate, Council of Scientific and Industrial Research, New Delhi-110 012.

ANONYMOUS (1988-89). Annual Report, Department of Horticulture. Narendra Deva University of Agriculture and Technology, Faizabad, U.P.

ANURADHA AGARWAL; RAM, H.Y.M. (1995). *In vitro* germination and micropropagation of water chestnut (*Trapa* sp.). *Aquatic Botany,* **51**(1/2): 135-146.

ARATHI, H.S.; GANESHAIAH, K.N.; SHAANKER, R.U.; HEGDE, S.G. (1996). Factors affecting embryo abortion in *Syzygium cuminii* (L.) Skeels (Myrtaceae). *International Journal of Plant Sciences,* **157**(1): 49-52.

ARIMA, S. (1994). [Studies on growth and yield performance of water chestnut (*Trapa bispinosa* Roxb.).] *Bulletin of the Faculty of Agriculture, Saga University,* No. **76**: 1-79.

ARIMA, S.; TANAKA, N.; MATSUMOTO, K. (1989). [Studies on growth and production of water chestnut. The effects of nitrogen application on the growth and yield of *Trapa bispinosa* Roxb.] *Report of the Kyushy Branch of the Crop Science Society of Japan,* No. **56**: 92-96.

ARSECULERATNE, S.N.; GUNATILAKA, A.A.L.; PANABOKKE, R.G. (1982 a). Studies on the toxicology of the palmyrah palm (*Borassus flabellifer* L.): Part I. A bioassay for the neurotoxin. *Journal of the National Science Council of Sri Lanka,* **10**(2): 269-275.

ARSECULERATNI, S.N.; GUNATILANKA, A.A.L.; PANABOKKE, R.G. (1982 b). Studies on the toxicology of the palyrah palm (*Borassus flabellifer* L.). II. Milk transfer to suckling rats. *Journal of the National Science Council of Sri Lanka,* **10**(2): 277-282.

ARUMUGAM, S.; RAO, M.V. (1996). *In vitro* production of plantlets from cotyledonary node cultures of *Aegle marmelos* (L.) Corr. *Advances in Plant Sciences,* **9**(2): 181-186.

ARUMUGAM, T.; SUTHANTHIRAPANDIAN, I.R.; DORAIPANDIAN, A. (1994). Effect of organic and inorganic manuring on neera and yield of palmyrah (*Borassus flabellifer* L.). *South Indian Horticulture,* **42**(4): 236-238.

ATTIRI, B.L.; SINGH, D.B. (1997). Composition of West Indian cherry (*Malpighia glabra* L.) at different stages of harvesting. *Journal of the Andaman Science Association,* **13**(1/2): 83-85.

AWAL, A. (1996). Influence of environment on polysaccharide contents of the seeds of *Borassus flabellifer* Linn. *Bangladesh Journal of Scientific and Industrial Research,* **31**(3): 69-80.

BAJPAYEE, K. K. (1997). Ethnobotany of *Phoenix* (Arecaceae). *Journal of Economic and Taxonomic Botany,* **21**(1): 155-157.

BALAKRISHNAN, K.; RAJENDRAN, C.; KILANDAIVELU, G. (2000). Differential responses of iron, magnesium and zinc deficiency on pigment composition, nutrient content, and photosynthetic activity in tropical fruit crops. *Photosynthetica,* **38**(3): 477-479.

BALASUBRAMANYAN, S.; BANGARUYSAMY, U. (1998). Maturity standard of aonla (*Emblica officinalis* Gaertn) under rainfed vertisol. *South Indian Horticulture,* **46**(3/6): 347-348.

BANDARA,   B.M.R.;   GUNATILAKE,   A.A.L.; SOTHEESWARAN, S.; WUERATNE, E.M.K.; RANASINGHE, M.A.S.K. (1989). Insecticidal properties and an active constituent of *Limonia acidissima. Journal of the National Science Council of Sri Lanka,* **17**(2): 237-239.

BANDOPADHYAY, D.P.; NATH, N.; PANDEY, H.S.; YADAV, L.P. (1983). Effect of type of rooting in stem cuttings of *Carissa* species. *Plant Science,* **12**: 44-45.

BANKAR, G.J. (1987). A note on Influence of gibberellic acid on seed germination and vigour of seedlings in Karonda (*Carissa carandas* L.). *Progressive Horticulture,* **19**(1-2): 90-92.

BARARE, D.; CHRETIEN, L. (1985). Floral anatomy of *Monstera deliciosa* (Araceae). *Canadian Journal of Botany,* **63**(8): 1428.

BARTHAKUR, N. N.; ARNOLD, N. P. (1989). Certain organic and inorganic constituents in bael (*Aegle mannelos* Correa) fruit. *Tropical Agriculture, U.K.,* **66**(1): 65-68.

BARTHAKUR, N.N.; ARNOLD, N.P. (1991). Chemical analysis of the emblic (*Phyllanthus emblica* L.) and its potential as food source. *Scientia Horticulturae,* **47**(1-2): 99-100.

BASKARAN, S.; BAI, M.K. (1998). Insect visitors of wood apple, *Limonia acidissima* L. (Rutaceae). *Insect Environment,* **4**(3): 88.

BEEL, E. (1998). Substrates for greenhouse plants cultivated with several irrigation systems. *Acta Horticulturae,* No. **221**: 315-326.

BERA, P.K.; MAZUMDAR, B.C. (1991). Ether induced ripening of carambola fruits. *Journal of National Botanical Society,* **45**: 71-73.

BERA, P.K.; MAZUMDAR, B.C. (1992). Comparative effect of spraying cashew trees with three auxun solutions on the qualitative constituents of their apples and kernels during developmental stages. *Indian Cashew Journal,* **24**(3): 6-10.

BEZERRA, J.E.F.; LEDERMAN, I.E.; FREITAS, E.V.de; SANITOS, V.F.dos (1999). [Effect of grafting method and rootstock age on propagation of Surinam cherry (*Eugenia uniflora* L.).] *Revista Brasileira* de *Fruticulturu,* **21**(3): 262-265.

BHAGAT, B.K.; JAIN, B.P.; SINGH, C. (1999). Success and survival of intergeneric grafts in guava (*Psidium guajava* L.). *Journal of Research, Birsa Agricultural University,* **11**(1): 79-81.

BHAKARE, H.A.; KHOTPAL, R.R.; KULKARNI, A.S. (1993). Lipid composition of *Withania somnifera, Phoenix sylvestris* and *Indigofera enualphylla* seeds of central India. *Journal of Food Science and Technology (Mysore),* **30**(5): 382-384.

BHANU VERMA; KANT, U. (1999). Propagation of *Emblica officinalis* Gaertn through tissue culture. *Advances in Plant Sciences,* **12**(1): 21-25.

BHARDWAJ, L.; MERILLON, J.M.; RAMAWAT, K.G. (1995). Changes in the composition of membrane lipids in relation to differentiation in *Aegle marmelos* callus cultures. *Plant Cell Tissue and Organ Culture,* **42**(1): 33-37.

BHARGAVA, B.S.; VARALAKSHMI, L.R.; KUMAR, B.P. (1997). New technique to standardize leaf sampling in phalsa (*Grewia asiatica* L.DC.) for nutritional diagnosis. *South Indian Horticulture,* **45**(1/2): 34-37.

BHATTACHARYYA, A.K.; ROY, G.C.; MAZUMDAR, B.C. (1988). De-greening of carambola fruits by steeping with brine water. *Science and Culture*, **54**: 373-375.

BHATTACHARYYA, A.K.; BERA, P.K.; ROY, G.C.; MAZUMDAR, B.C. (1989). Studies on nut cracking problem of cashewnut (*A. occidentale* L.) in the southern part of West Bengal. *Cashew Bulletin*, **26**(8): 1-5.

BHATTACHARYYA, A.K.; ALI, S.L.; MAZUMDAR, B.C. (1990). Qualitative constituents of longan (*Euphoria longan* Steud) fruits during their developmental stages. The *Andhra Agriculture Journal*, **37**(2): 192-194.

BHATTACHARYYA, A.K.; ALI, S.L.; MAZUMDAR, B.C. (1992). Qualitative constituents of some minor fruits during their developmental stages. *Indian Agriculturist*, **36**(2): 125-126.

BHATTACHARYYA, A.K.; MAZUMDAR, B.C. (1993). Effect of spraying cashew (*A. occidentale* L.) tress with penicillin solutions on qualitative constituents of their apples and kernels. *Journal of the National Botanical Society*, **47**: 31-36.

BHATTACHARYA, S.; DAS, S.; MUKHERJEE, K.K. (1993). Biochemical studies on palm pollen. *Grana*, **32**(2): 123-127.

BHAUMIK, P.K.; GUHA, K.P.; BISWAS, G.K.; MUKHERJEE, B. (1987). Flacourtin, a phenolic glucoside ester from *Flacourtia indica*. *Phytochemistry*, **26**(11): 3090-3091.

BICALCO, B.; REZENDO, C.M. (2001). Volatile compounds of cashew apple (*A. occidentale* L.). *Zeits. Natur. Section C, Biosciences*. **56**(1/2): 35-39.

BORA, A.; DAS, R.P. (1998). Role of indole butyric acid (IBA) on vegetative growth of rooted cuttings of some minor fruits of Assam. *Journal of the Agricultural Science Society of North East India*, **11**(2): 197-201.

BRASIL, E.C.; SILLVA, A.M.B.; MÜLLER, C.H.; SILVA, G.R. da (1999). [Effect of nitrogen and potassium fertilization and of limestone on the development of Barbados cherry seedlings.] *Revista Brasileira de Fruticultura*, **21**(1): 52-56.

CAMPELO, C.R. (1988). [Contribution to the study of medicinal plants in the Estado de Alagoas, V.) *Acta Amazonica (Suplemento)*, **18**(1-2): 305-312.

CARTAGENA VALENZUELA, J.R.; BARRETO OSORIO, J.D. (1998). [Effect of gibberellic acid and sowing method on seed germination and seedling grown of bullock's heart (*Annona reticulata* L.).] *Agronomia Medellin*, **51**(2): 235-244.

CARVALHO, R.I.N. de (1998). [Variability in young Barbados cherry plants propagated from seed.] *Agropecuaria Catarinense*, **11**(1): 16-18.

CHAKRABORTY, D.; MAHAPATRA, P.K.; NAG CHAUDHURI, A.K. (1986). A neuropsycho-pharmacological study of *Syzygium cumini*. *Planta Medica*, No. **2**: 139-143.

CHANDRA, G. (1985). Investigations on essential oils and isolates of potential value at HBTI, Kanpur, *Indian Perfumer*, **29**(1/2): 23-30.

CHANG CHINGHUI; YU TUNGHSI; LIN LIYUN; CHANG CHIYUE (1998). [Studies on the important volatile flavor compounds of fresh and dried longan fruits.] *Journal of the Chinese Agricultural Chemical Society*, **36**(5): 521-532.

CHANTARANOTHAI, P.; PARNELL, J.A.N. (1994). The breeding biology of some Thai *Syzygium* species. *Tropical Ecology*, **35**(2): 199-208.

CHAO CHENGNAN; CHANG JERWAY; YEN CHUNGRUEY (1997). [Variety improvement of longan (*Dimocarpus longan* Lour.) in Taiwan.] In: *Proceedings of a symposium on enhancing competitiveness of fruit industry, Taichung, Taiwan, 20-21 March 1997 (edited by Chen Yungwu and Chang LinRen).*

CHATTOPADHYAY, P.K.; AMITAVA GHOSH (1996). Mineral composition of inflorescence and fruits at different stages of growth in rose apple. *Orissa Journal of Horticulture*, **20**(2): 28-31.

CHATTOPADHYAY, P.K.; MAHANTA, S.K. (1989). Bael: media requirements for seed germination and seedling establishment. *Indian Horticulture*, **36**(3): 27.

CHAUDHURI, A.K.N.; PAL, S.; GOMES, A.; BHATTACHARYA, S. (1990). Anti-inflammatory and related actions of *Syzygium cuminii* seed extract. *Phytotherapy Research*, **4**(1): 5-10.

CHEN JINGYING; CHEN JING YAO; CHEN XI (1999). [Shoot apex culture of longan (*Dimocarpus longan* Lour.) trees.] *Journal of Tropical and Subtropical Botany*, **7**(2): 159-164.

CHEN QIXUAN; CHEN LEL JIN; YU DESHENG; LI CHUN (2000). [Qingshan O, a promising late variety of longan.] *South China Fruits*, (2000) **29**(1): 22.

CHOU HUTNA; O. SHAUMEI; NEE CHENGCHU. (2000). (Studies on the changes in composition of Taiwan kiwifruits during fruit storage.] *Journal of the Chinese Society for Horticultural Science*, **46**(2): 157-172.

CHOVATIA, R.S.; SINGH, S.P. (2000 a). Effect of time on budding and grafting success in *jamun* (*Syzgium cumini* Skeel). *Indian Journal of Horticulture*, **57**(3): 255-258.

CHOVATIA, R.S.; SINGH, S.P. (2000 b). Success of air-layering in custard apple (*Annona squamosa* L.) as influenced by ringing of shoots and growth regulators. *Orissa Journal of Horticulture*, **28**(2): 61-65.

CHOWDHURY, M.K. (1992). *Kendbona Eco-Development Project:* A novel approach to wasteland reclamation. *Indian Forester*, **118**(2): 879-886.

COZZA, R.; GALANT, I.G.; BITONTI, M.B.; INNOCENTI, A.M. (1994). Effect of storage at low temperature on the germination of the waterchestnut (*Trapa natans* L.). *Phyton* (*Horn*), **34**(2): 315-320.

CRUZ, V.D'.A.; D'ARCE, P.G.; CASTILHO, V.M.; LIMA, V.A. de; CRUZ, R.; GODINHO, P.H. (1995). [Changes in the ascorbic acid content of acerolas (*Malpighia glabra* L.) as a function of maturity stage and storage temperature.] *Arquivos de Biologia e Tecnologia*, **25**(2): 331-337.

DAMASCENO JUNIOR, J.A.; BEZERRA, F.C. (2002). [Quality of irrigated early dwarf cashew peduncles in different management and spacing systems. *J. Revista Brasileira de Fruitcultura*, **24**(1): 258-262.

DARBARA SINGH; JINDAL, P.C. (1979). Seed germination studies in *phalsa* (*Grewia asiatica* L.). *Haryana Journal of Agriculture Science*, **8**(1/2): 47-48.

DAULATABAD, C.M.J.D.; MIRAJAKAR, A.M.; HOSAMANI, K.M.; MULLA, G.M.M. (1988). Epoxy and cyclopropenoid fatty acids in *Syzygium cumini* seed oil. *Journal of the Science of Food and Agriculture*, **43**(1): 91-94.

DE RUPA; MITRA JAYA; MAJUMDER, K.; MAZUMDAR, B.C. (1993). Laminar nitrogen, phosphorus and potassium content during winter to summer seasons in some fruits grown in West Bengal. *Indian Biologist*, **28**(2): 37-39.

DHANDAR, D.G.; BHARGAVA, B.S. (1993). Leaf sampling technique for nutritional diagnosis in custard apple. *Indian Journal of Horticulture*, **50**(1): 1-4.

DHANKHAR, D.S.; SANTOSH KUMAR. (1996). Effect of bio-regulator on seed germination and seedling growth in Aonla (*Phyllanthus emblica* Linn.) cv. Anand-2. *Recent Horticulture*, **3**(1): 45-48.

DHANKHAR, D.S.; SINGH, M. (1996). Seed germination and seedling growth in aonla (*Phyllanthus emblica* Linn.) as influenced by gibberellic acid and thiourea. *Crop Research (Hisar)*, **12**(3): 363-366.

EGBEKUN, M.K.; OTIRI, A.O. (1999). Changes in ascorbic acid contents in oranges and cashew apples with maturity. *Ecology of Food and Nutrition*, **38**(3): 275-284.

EL SIDDIG, K.; LÜDDERS, P.; EHERT, G.; ADIKU, S.G.K. (1998). Response of rose apple (*Eugenia jambos* L.) to water and nitrogen, supply. *Angewandte Botanik*, **72**(5/6): 203-206.

FERRO, E.; SCHININI, A.; MALDONADO, M.; ROSNER, J.; SCHMEDA HIRASCHMANN, G. (1988). *Eugenia uniflora* leaf extract and lipid metabolism in *Cobs apella* monkeys. *Journal of Ethnopharmacology*, **24**(2-3): 321-325.

FIGUEIREDO, R.W.; LAJOLO, F.M.; ALVES, R.E.; FIGUEIRAS, H.A.C.; ARAUJO, N.C.C. (2001). [Change of firmness, pectins and pectolytic enzymes during the development and maturation of fruits of early dwarf cashew (*A. occidentale* L. var. *nanum*) CCP-76.] *Proceedings of the Interamerican Society for Tropical Horticulture*, **43**: 82-86.

FIGUEIREDO, R.W. de; LAJOLO, F.M.; AILVES, R.E.; FILGUEIRAS, H.A.C. (2002). Physical-chemical changes in early dwarf cashew pseudofruits during development and maturation. *Food Chemistry*, **77**(3): 343-347.

FILGUEIRAS, H.A.C.; ALVERS R.E.; MOSCA, J.L.; MENEZES, J.B. (1999). Cashew apple for fresh consumption. *Acta Horticulture*, No. **485**: 155-160.

FISCHER, G.; LÜDDERS, P. (1997). Developmental changes on carbohydrates in cape gooseberry (*Physalis peruviana* L.) fruits in relation to the calyx and the leaves. *Agronomfa Colombiana*, **14**(2): 95-107.

FISCHER, G.; LÜDDERS, P.; GALLO, F. (1997). [Quality changes of the cape gooseberry fruit during its ripening.] *Erwerbsotbau*, **39**(5): 153-156.

FISCHER, G.; MARTINEZ, O. (1998). [Quality and maturity of cape gooseberry (*Physalis peruviana* L.) in relation to fruit colour.] *Agronomy Colombiana*, **16**(1/3): 35-39.

FOURIE, C. (1988). [Success with growing-point grafting for citrus.) *Information Bulletin, Citrus and Sub-tropical Fruit Research Institute, South Africa*, No. **187**: 1-2.

GALHO, A.S.; LOPES, N.F.; RASEIRA, A.; BACARIN, M.A. (2000). [Growth of *Psidium cattleyanum* Sabina fruits.] *Revista Brasileria de Fruticultura*, **22**(2): 223-225.

GARNDE, V.K.; JOSHI, G.D.; MAGDUM, M.B.; WASKAR, D.P. (1998 a). Studies on physical changes during growth and development of *jamun* (*Syzygium cuminii* Skeels) fruit. *Agricultural Science Digest* (Karnal), **18**(3): 206-208.

GENUA, J.M.; HILLSON, C.J. (1985). The occurrence, type and location of calcium oxalate crystals in the leaves of fourteen species of Araceae. *Annals of Botany*, **56**(3): 351-361.

GHOLAP, S.V.; DOD, V.N.; BHARAD, S.G.; WANKAR, A.M. (2000 a). Studies on vegetative propagation in bullock's heart (*Annona reticulata* L.). *Crop Research* (*Hisar*), **20**(1): 158-159.

GHOLAP, S.V.; DOD, V.N.; BHUYAR, S.A.; BHARAD, S.G. (2000 b). Effect of plant growth regulators on seed germination and seeding growth in aonla (*Phyllanthus emblica* L.) under climatic condition of Akola. *Crop Research* (*Hisar*), **20**(3): 546-548.

GHOSH, A.K.; NAGAR, P.K.; SIRCAR, P.K. (1985). Endogenous gibberellins in developing fruits of *Ziziphus mauritiana* Lam. and *Aegle marmelos* Correa. *Journal of Plant Physiology*, **120**(5): 381-388.

GHOSH, P.; BANDYOPADHYAY, A.K.; THAKUR, S.; AKIHISA, T.; SHIMIZU, N.; TAMURA, T.; MATSUMOTO, T. (1989). Acidissimin, a new tyramine derivative from the fruit of *Limonia acidissima*. *Journal of Natural Products*, **52**(6): 1323-1326.

GOMES, J.; HOSSAIN G.; AWAL, A.; HAQ, Q.N. (1986). Polysaccharide components in the pulp and the seeds of *Eugenia jambolana* L. (*Syzygium cumini*. *Bangladesh Journal of Scientific and Industrial Research* (3/4): 186-194.

GOMEZ, P.; REYNES, M.; DORNIER, M.; HEBRRT, J.P. (1999). [The West Indian cherry: a valuable natural source of vitamin C.] *Fruits (Paris)*, **54**(4): 247-260.

GONZAGA NETO, L.; AMORAL, M.G.DO; SAUERESSIG, M.E. (1996). [Grafting and budding of Barbados cherry under semi-shade conditions.] *Pesquisa Agropecudria Brasileira*, **31**(9): 635-638.

GOPALAN, C.; RAMA SASTRI, B.V.; BALA SUBRAMANIAN, S.C. [Revised & Updated by NARASINGA RAO, B.S.; DEOSTHALE, Y.G.; PANT, K.C.] (1993). *Nutritive Value of Indian Foods*. National Institute of Nutrition, Indian Council of Medical Research, Hyderabad-500 007.

GUADARRAMA, A. (1984). [Some chemical changes during ripening of Barbados cherry (*Malpighia punicifolia*) fruits.] *Revista de la Facultad de Agronomfa. Universidad Central de Venezuela*, **13**(1/4): 111-128.

GUHA, GOPA; MAZUMDAR, B.C. (1996). Vitamin C content in different layers of fruits. *Applied Nutrition*, **21**(2): 40.

GUHA BAKSHI, D.N.; SENSARMA, P.; PAL, D.C. (1999). *A Lexicon of Medicinal Plants in India*. Naya Prokash, Calcutta-700 006.

GURBACHAN SINGH; BHIST, J.K.; SINGH, N.T.; DAGAR, J.C. (1998). Evaluation of fruit plants for sodicity tolerance. *Range Management & Agroforestry,* **19**(1): 87-91.

HANDIQUE, P.J.; BHATTACHARJEE, S. (2000). *In vitro* shoot proliferation of wood apple (*Feronia elephantum Correa*). *Advances in Plant Sciences,* **13**(1): 241-243.

HASNAIN, ALI; AU, R. (1988). Protein and amino acids of *Grewia asiatica. Pakistan Journal of Scientific and Industrial Research,* **31**(11): 777-779.

HAYES, W.B. (1970). *Fruit Growing in India.* Kitabistan, Allahabad, India.

HAZARIKA, B.N.; NAGARAJU, V.; PARTHASARATHY, V.A. (1996). Morphogenetic response of bael (*Aegle marmelos* L.) microshoots to benzyl amino purine. *Annals of Plant Physiology* **10**(1): 40-44.

HAZARIKA, B.N.; NAGARAJU, V.; PARTHASARATHY, V.A. (1998). *Ex vitro* acclimatisation of microshoots of *Aegle marmelos* L. *International Journal of Tropical Agriculture,* **14**(1/4): 251-253.

HEINZE, W.; MIDASCH, M. (1991). [Photoperiodic reaction of *Physalis peruviana* L.] *Gartenbauwissenschaft,* **56**(6): 262-264.

HEREDIA, J.B.; SILLER, J.H.; BARZ, M.A.; ARAIZA, E.; PORTILLO, T.; GARCIA, R.; MUY, M. D. (1998). [Changes in the quality and content of carbohydrates in tropical and sub-tropical fruits at the supermarket level.] *Proceedings of the Interamerican Society for Tropical Horticulture,* **41**: 104-109.

HORE, J.K.; SEN, S.K. (1985). Effect of seed treatments on longevity of bael (*Aegle marmelos* C) seeds. *Haryana Journal of Horticultural Sciences,* **14**(3/4): 204-210.

HORE, J.K.; SEN, S.L. (1994). Effect of non-auxinic compounds and IBA on root formation in stem cuttings of rose apple (*Syzygium jambos* Alston). *Crop Research (Hisar)*, **7**(1): 44-48.

HORE, J.K.; SEN, S.K. (1995). Effect of seed treatment on germination, seedling growth and longevity of bael (*Aegle marmelos* C) seeds. *Agricultural Science Digest (Karnal)*, **15**(1/2): 26-30.

HOSSAIN, M.; BISWAS, B.K.; KARIM, M.R.; RAHMAN, S.; ISLAM, R.; JOARDER, O.I. (1994 a). *In vitro* organogenesis of elephant apple (*Feronia limonia*). *Plant Cell, Tissue and Organ Culture*, **39**(3): 265-268.

HOSSAIN, M.; ISLAM, R.; KARIM, M.R.; RAHMAN, S.M.; JOARDER, O.I. (1994 b). Production of plantlets from *Aegle marmelos* nucellar callus. *Plant Cell Reports*, **13**(10): 570-573.

HAUNG JINSONG; XU XIUDAN; ZHANG SHAOQUAN; ZHU TANGMING; LI CHUN (1996). [A new, large-fruited, late ripening longan cultivar-Qingke Baoyuan.] *China Fruits*, No. **4**: 4-5, 7.

HUSSAIN, M.D.; HUSSAIN, M.I.; ALAM, M. (1992). Juice harvesting from date and palmyra palm tree in Bangladesh. *Indian Journal of Agricultural Engineering*, **2**(1): 17-24.

HSU YUMEI; TSENG MENQJAIU; LIN CHINHO. (1999). The fluctuation of carbohydrates and nitrogen compounds in flooded wax-apple trees. *Botanical Bulletin of Academia Sinica*, **40**(3): 193-198.

ISLAM, B.N. (1986). Use of some extracts from Mellaceae and Annonaceae for control of rice hispa. *Dicladispa armigera*, and, the pulse beetle, *Callosobruchus chinensis*. [In: *Natural Pesticides from the Neem Tree (Azadirachta indica A. Juss) and Other Tropical Plants.*] *Proceedings of the 3rd International Neem Conference, Nairobi, Kenya, 10-15 July 1986* [edited by Schmutterer, H.; Ascher, K.R.S.].

ISLAM, R.; AHAD, A.; RAHMAN, M.H.; HOSSAIN, M.; JOARDER, O.I. (1996 a). High frequency adventitious plant regeneration from radicle explants of *Aegle marmelos* Corr. *Pakistan Journal of Botany*, **28**(2): 203-206.

ISLAM, R.; HOSSAIN, M.; JOARDER, O.I.; KARIM, M.R. (1993). Adventitious shoot formation on excised leaf explants of *in vitro* grown seedlings of *Aegle marmelos* Corr. *Journal of Horticultural Science*, **68**(4): 495-498.

ISLAM, R.; HOSSAIN, M.; REZA, M.A.; MAMUN, A.N.K.; JOARDER., O.I. (1996 b). Adventitious shoot regeneration from root tips of intact seedlings of *Aegle marmelos*. *Joumal of Horticultural Science*, **71**(6): 995-1000.

ISLAM, R.; KARIM, M.R.; RAHMAN, S.M.; HOSSAIN, M.; JOARDER, O.I. (1994). Plant regeneration from excised cotyledon of *Aegle marmelos* Corr. *Pakistan Journal of Botany*, **26**(2): 393-396.

JAISWAL, H.R.; MISRA, K.K.; VIJAI SINGH (1999). Clonal variations in growth and yield of bael (*Aegle marmelos* Correa). *Scientific Horticulture*, **6**: 31-36.

JANA, S.; MAZUMDAR, B.C. (1977). Physico-chemical analyses of water-chestnut (*Trapa natans*). *Science and Culture*, **43**: 361-362.

JAUHARI, O.S.; SINGH, R.D. (1971). *Bael*–Valuable fruit. *Indian Horticulture*, **16**: 9-10.

JAUHARI, O.S.; SINGH, R.D.; AWASTHI, R.K. (1969). Survey of some important varieties of bael (*Aegle marmelos* Cor.). *Punjab Horticultural Journal*, **9**: 48-53.

JAYANT KUMAR; PARMAR, C. (2000). Standardization of sexual and asexual propagation techniques for some wild fruits of sub-Himalayan region *Indian Forester*, **26**(8): 870-873.

KADER, S.A.; BINDU, K.R.; CHACKO, K.C. (2000). Polyembryony in *Syzygium cumini* (L.) Skeels and in *Vateria indica* L. *Indian Forester*, **126**(12): 1353-1356.

KADONO, Y.; SCHNEIDER, E.L. (1986). Floral biology of *Trapa natans* var. *japonica*. *Botanical Magazine,* **99**(1056): 435-439.

KANCHAN KUMAR SRIVASTAVA; SINGH, H.K. (2000). Floral biology of bael (*Aegle marmelos*) cultivars. *Indian Journal of Agricultural Science,* **70**(11): 797-798.

KAMALUDDIN, M.; NATH, T.K.; JASHIMUDDIN, M. (1998). Indigenous practice of khejur palm (*Phoenix sylvestris*) husbandry in rural Bangladesh. *Journal of Tropical Forest Science,* **10**(3): 357-366.

KANDIAH, S.; KOKUDATHASAN, S. (1987). EDTA-stimulation of sap (toddy) flow from inflorescence during tapping in palmyrah palm (*Borassus flabellifer* L.). *Vinganam Journal of Science,* **2**(1-2): 58-61.

KANDIAH, S.; MAHENDRAN, S. (1986). A new method for culturing palmyrah palm *Borassus flabellifer* L. *Vingnanam Journal of Science,* **1**(1): 40-43.

KANT, U.; BHANU VERMA (1997). Propagation of *Emblica officinalis* Gaertn. through tissue culture. [In: *Plant Tissue Culture and Biotechnology: Emerging Trends.*] *Proceedings of a Symposium held at Hyderabad, India, 29-31 January 1997 [edited by Kishor P.B.K.].* Hyderabad, India.

KANTHARAJAH, A.S.; RICHARDS, G.D.; DODD, W.A. (1992). Roots as a source of explants for the successful micropropagation of carambola (*Averrhoa carambola* L.). *Scientia Horticulturae,* **51**(1-2): 169-177.

KAPOOR, N.; BEDI, K.L.; TIKOO, M.K.; KAPOOR, S.K.; SARIN, Y.K. (1990). Some lesser known tree-borne oilseeds of India. *Journal of the Oil Technologists' Association of India,* **22**(3): 65-66.

KARALE, A.R.; KESKAR, B.G.; DHAWALE, B.C.; SHETE, M.B.; KALE, P.N. (1991). Flowering, sex expression and sex ratio in Banarasi aonla seedling trees. *Journal of Maharashtra Agricultural Universities,* **16**(2): 270-271.

KARALE, A.R.; KESKAR, B.G.; SHETE, M.B.; DHAWALE, B.C.; KALE, P.N.; CHOUDHARI, K.G. (1989). Seedling selection in *karonda* (*Carissa carandas* L.). *Maharashtra Journal of Horticulture.* **4**: 125-129.

KARUNANAYAKE, E.H., WELTHINDA, J.; SIRIMANEE, S.; SINNADORAI, R. (1984). Oral hypoglycaemic activity of some medicinal plants of Sri Lanka. *Journal of Ethnopharmacology,* **11**(2) 223-231.

KE GUANWU; WANG CHANGCHUN; TANG ZIFA (1994). [Palynological studies on the origin of longan cultivation.] *Acta Horticulturae Sinica,* **21**(4): 323-328.

KESKAR, B.C.; DHAWALE, B.C.; KARALE, A.R.; KALE, P.N. (1991). Propagation techniques in *aonla. Journal of Maharashtra Agricultural University,* **6**(2): 282.

KHANNA, R.K.; SUBHASH CHANDRA. (1996). Forest/ domestic as a source of natural dyes. *Journal of Economic and Taxonomic Botany,* **20**(2): 497-500.

KHATTAB, M.; KAMEL, H.; YACOUT, M. (1987). Nitrogen and potassium nutrition of *Monstera deliciosa,* Liebm. *Alexandria Journal of Agricultural Research,* **32**(2): 277-288.

KNIGHT, R.J. Jr.; CAMPBELL, C.W. (1993). Pollination requirements for successful fruiting of tropical fruit species. [In: XXXIX *Annual Meeting of the Interamerican Society for Tropical Horticulture. Santo Domingo, Dominican Republic, 22-27Aug., 1993 (edited by Campbell, R.J.)] Proceedings of the Interamerican Society for Tropical Horticulture,* **37**: 167-170.

KOSUGE, T.; YOKOTA, M.; SUGIYAMA, K.; OKAMOTO, A.; SAITO, M.; YAMAMOTO, T. (1986). Studies on Chinese medicines used for cancer. III. Cytotoxic constituent against HeLa cells in the fruit of *Trapa bispinosa* Roxb. *Yakugaku Zasshi* **106**(2): 183-185.

KUMAR, D.P.; KHAN, M.M.; MELANTA, K.R. (1996). Effect of nutrition and growth regulators on apple characters and yield in cashew (*Anacardium occidentale* L.). *Cashew,* **10**(2): 17-24.

KUMAR, P.H.; SREEDHARAN, C. (1987). Correlation studies between leaf nutrients and fruit quality characters in cashew (*Anacardium occidentale* L.). *Indian Cashew Journal,* **18**(3): 15-16.

KUMRI, M.; MAJUMDER, K.; MAZUMDAR, B.C. (1997). Qualitative constituents of some types of bael (*Aegle marmelos* Cor.) fruits growing in West Bengal. *Indian Biologist,* **29**(2): 52-54.

LI JI HONG; SHAO HANSHUANG; HUANG GUIXIU; FU SHAOP (1999).[Experiment of fast propagation of carambola *in vitro.*] *China Fruits,* No. **4**: 37.

LING JINGTIAN; IWAMASA, M. (1997). Plant regeneration from embyogenic calli of six *Citrus* related genera. *Plant Cell, Tissue and Organ Culture,* **49**(2): 145-148.

LITZ, R. E. (1984). *In vitro* responses of adventitious embryos of two polyembryonic *Eugenia* species. *HortScience,* **19**(5): 720-722.

LIYUQIAO; ZHOU CUIPING (1997). [Study on the technique of tissue culture for *Monstera deliciosa* Liebn.] *Journal of Jiangsu Forestry Science and Technology,* **24**(1): 42-45.

LU, Z.Q.; ZHU, J.; ZHU, S.; CHEN, Z.D. (1984). [Preliminary studies on the beetle (*Galeruceila birmanica Jacoby*) an insect pest of waterchestnut and watershield.) *Scientia Agricultura Sinica,* No. **5**: 73-76.

MACLEOD, J.K.; MOELLER, P.D.R.; BANDARA, B.M.R.; GUNATILAKA, A.A.L.; WIJERATNE, E.M.K. (1989). Acidissimin, a new limonoid from *Limonia acidissima.* *Journal of Natural Products,* **52**(4): 882-885.

MADHAVA RAO, V.N. (1969). *Cashewnut Cultivation in India*. Ministry of Food, Agriculture, C.D. and Co-operation, New Delhi.

MAGALHAES, L.M.F.; OLIVEIRA, D.de; OHASHI, O.S. (1999). [Effect of pollination on acerola fruit set in the Amazon region.] *Brasileina de Fruitcultura*, **21**(1): 95-97.

MAIKUP, M.; MAJUMDER, K.; MAZUMDAR, B.C. (1997). Important constituents in apple and kernel of tree types of cashew grown in southern part of West Bengal. *South Indian Horticulture*, **45** (5 and 6): 299-300.

MAITI, C.S.; NATH, A.; SEN, S.K. (1999 b). Studies on the propagation of bael (*Aegle marmelos* Correa) by different grafting methods in West Bengal. *Journal of Applied Horticulture, Lucknow*, **1**(2): 131-132.

MAITI, C.S.; NATH, A.; SEN, S.K.; (1999 b). Physico-chemical analysis of fruits of bael (*Aegle marmelos* Correa) cultivars of West Bengal. *Journal of Interacademicia*, **3**(2): 168-171.

MAJUMDER, K. (1990). *Involvement of Ethylene on Changing Pattern of Pectic Polysaccharides and Related Enzymes in Developing Fruits of Physalis peruviana, L.* Ph.D. thesis submitted to the Calcutta University, under the supervision of research by Dr. B. C. Mazumdar.

MAJUMDAR, K.; MAZUMDAR, B.C. (2001 a). Effects of auxin and gibberellin on pectic substances and their degrading enzymes in developing fruits of cape-gooseberry (*Physalis peruviana* L.). *Journal of Horticultural Science and Biotechnology*, **76**(3): 276-279.

MAJUMDER, K.; MAZUMDAR, B.C. (2001 b). A note on pectin content in some under-utilized minor fruits of India. *Science* and *Culture*, **67**(3-4): 125.

MAJUMDER, K.; MAZUMDAR, B.C. (2002). Changes of pectic substances in developing fruits of cape-gooseberry (*Physalis peruviana* L.) in relation to the enzymic activity and evolution of ethylene. *Scientia Horticulturae,* **96**: 91-101.

MANDAL, U. (1997).*Quantitative Changes of Important Constituents in Some Minor Fruits During Development and Improvement of the Constituents in a Minor Fruit (Carissa carandas L.) of Agro-industrial Importance.* Ph.D. thesis submitted to the Calcutta University, under the supervision of research by Dr. B.C. Mazumdar.

MANDAL, U.; MAJUMDER, K.; MAZUMDAR, B. C. (1997). Comparative study of some qualitative constituents of water-chestnuts. *Indian Biologist,* **19**(2): 37-38.

MANDAL, U.; MAZUMDAR, B. C. (1995). Physico-chemical constituents of some tropical and sub-tropical minor fruits of India. *Proceedings of the National Symposium on Sustainable Agriculture in Sub-humid Zone, Visva-Bharati, Sriniketan - 731 236, W. Bengal, India.*

MANDAL, U.; MAZUMDAR, B.C. (1997). Qualitative characteristics of some summer season tropical and sub-tropical minor fruits grown in West Bengal. *Indian Agriculturist,* **41**: 291-298.

MANDAL, U.; MAZUMDAR, B.C. (1998). Qualitative constituents in some tropical and sub-tropical minor fruits during development. *Indian Biologist,* **30**(1): 70-79.

MANDAL, U.; MAZUMDAR, B. C. (2000). Qualitative improvement of some minor fruits by foliar application of plant nutrients and growth regulants. *Indian Agriculturist,* **44**(1 and 2): 93-96.

MANDAL, U.; SINHA ROY, S.; MAZUMDAR, B.C. (1992). A recently developed agro-industry in the southern suburb of Calcutta city utilizing a bramble fruit. *Indian Journal of Landscape Systems Ecological Studies,* **15**(1): 100-102.

MARCELINO PONCE, J. (1986). [Performance of *Annona reticulata* grafted on various rootstocks.] *Proceedings of the Tropical Region, American Society for Horticultural Science,* **23**: 119-121.

MARLER, T.E.; MICKERBART, M.V. (1992). Application of GA_{4+7} to stem enhances carambola seedling growth. *Hort. Science,* **27**(2): 122-123.

MARTINS, A.B.G.; ANTUNES, E.C. (2000). [Propagation of rose apples (*Syzygium jambos* L. Alston) by air layering.] *Revista Brasileira de Fruticultura,* **22**(2): 205-207.

MAZUMDAR, B.C. (1975). Physico-chemical analyses of some types of bael (*Aegle marmelos* Correa) fruits growing in West Bengal. *Indian Agriculturist,* **19**(3): 295-298.

MAZUMDAR, B.C. (1979 a). Cape-gooseberry – the jam fruit-cultivation in India. *World Crops* (London), **31**(1): 19, 23.

MAZUMDAR, B.C. (1979 b). Fruit analysis of two *Syzygium* species. *Plant Science,* **11**: 100.

MAZUMDAR, B.C. (1982). [Less known water-chestnuts.] *Basundhara* (Government of West Bengal), **34**: 21-24.

MAZUMDAR, B.C. (1983). [Cultivation of bael.] *Basundhara* (Government of West Bengal), **35**: 25-27.

MAZUMDAR, B.C. (1984). [Palmyra and wild date in afforestation.] *Dhanadhanney* (Ministry of Information and Broadcasting, Government of India), **16**(6): 20-21.

MAZUMDAR, B.C. (1985). Water-chestnut – the aquatic fruit: Cultivation in India. *World Crops* (London). **37**: 42-44.

MAZUMDAR, B.C. (1986). Qualitative behaviour of different types of emblica (*Phyllanthus emblica* L.) fruits and loss of their vitamin C by steeping with brine water. *Indian Biologist,* **18**(1): 34-36.

MAZUMDAR, B.C. (1989). A review on anti-fertility properties of higher plants. *Indian Journal of Lanscape Systems and Ecological Studies*, **12**(2): 53-60.

MAZUMDAR, B.C. (1990). Polyembryony in seeds of some fruit trees. *The Andhra Agricultural Journal*, **37**(3): 311-312.

MAZUMDAR, B.C. (2003). *Principles and Methods of Orchard Establishment in India.* Daya Publishing House, 4762-63/23, Ansari Road, Darya Ganj, New Delhi-110 002.

MAZUMDAR, B.C.; JANA, S. (1977). Physico-chemical analysis of water-chestnut (*Trapa bispinosa*) fruits. *Science and Culture*, **43**: 361-362.

MAZUMDAR, B.C.; MUKHOPADDHYA, P. (2003). [*Preserved Products from Fruits – Scientific Methods.*] Sri Bhumi Publishing Co., 79, Mahatma Gandhi Road, Kolkata-700 009.

MEHRA, K.L.; ARORA, R.K. (1982). *Plant Genetic Resources of India: Their Diversity and Conservation.* National Bureau of Plant Genetic Resources, New Delhi.

MEZOUITA, P.C.; VIGOA, Y.G. (2000). [The aceroda: marginal fruit from America with a high level of ascorbic acid.] *Alimentaria*, **37**: 413-425.

MICHELINI, S.; CHINNERY, L.E. (1988). The use of plant growth regulators and irrigation to control flowering of the acerola or Barbados cherry, *Malpighia glabra* L. *Proceedings of the Interamerican Society for Tropical Horticulture*, **32**: 65-73.

MILLER, W.R.; McDONALD, R.E. (1997). Carambola quality after ethylene and cold treatments and storage. *HortScience*, **32**(5): 897-899.

MISRA, K.K.; JAISWAL. H.R. (1993). Effect of growth regulators on rooting of stool layers of Karonda (*Carissa carandas* L.). *Indian Journal of Forestry*, **16**(2): 181-182.

MISRA, K.K.; JAISWAL, H.R. (1998). Effect of foliar sprays of gibberellic acid on the growth of Karonda (*Carissa carandas* L.) seedlings. *Indian Journal of Foresrty*, **21**(1): 70-71.

MISRA, K.K.; RAJESH SINGH; JAISWAL, H.R. (2000). Performance of bael (*Aegle marmelos*) genotypes under foot-hills region of Uttar Pradesh. *Indian Journal of Agricultural Sciences*, **70**(10): 682-683.

MISRA, K.K.; SINGH, R. (1990). Effect of growth regulators on rooting and survival of air layers of Karonda (*Carissa carandas* L.). *Annals of Agricultural Research*, **11**(2): 208-210.

MISRA, R.S.; BAJPAI, P.N. (1975). Studies on the floral biology of jamun [*Syzygium cumini* (L.) Skeels.) *Indian Journal of Horticulture*, **32**: 15-24.

MORTENSEN, L.M. (1991). The effect of air temperature on the growth of foliage plants. *Norwegian Journal of Agricultural Sciences*, **5**(3): 289-294.

MOSCHINI, F.; TOSI, D.; GRAIFENBERG, A. (1985). [The effect of type of fertilization on the raising of *Monstera deliciosa* Leib. and *Ficus benjamina* L. in hydroculture.] *Rivista della Ortoflorofrulticoltura Italiana*, **69**(6): 397-411.

MOTI SINGH; GAUTAM, R.K.S.; PRASAD, Y.; SINGH, A.K. (1986). Studies on the effect of plant regulators and their different concentrations on yield and quality of *phalsa* (*Grewia subinaequalis* D.C.). *Haryana Journal of Horticultural Sciences*, **15**(3/4): 196-199.

MOURA, C.F.H.; ALVES, R.E.; FILGUEIRAS, H.A.C.; INNECO, R.; PINTO, S.A.A. (2000). *Proceedings of the Interamerican Society for Tropical Horticulture*, **42**: 119-123.

MOURA, C.F.H.; ALVES, R.E.; INNECO, R.; FILGUEIRAS, H.A.C.; MOSCA, J.L.; PINTO, S.A.A. (2001). [Physical characteristics of cashew apples for fresh fruit market.] *Revista Brasileira de Fruiticultura*, **23**(3): 537-540.

MUKAI, H.; UTSUNOMIYA, N.; SUGIURA, A. (1989). [The effects of temperature on the fruit maturation and quality of the two types of strawberry guava, *Psidium cattleianum* Sabine and *Psidium cattleianum* Sabine var. *lucidum* Degener.] *Japanese Journal of Tropical Agriculture*, **33**(4): 243-247.

MUKHERJEE, S.K.; RAO, D.P.; CHAKLADAR, B.P.; CHATTERJEE, B.K. (1986). Effect of growth regulators, invigoration and etiolation on rooting of air layers of bael (*Aegle marmelos* Correa). *Indian Journal of Horticulture*, **43**(1/2): 9-12.

MUKHOPADHYAA, D.P. (1982). Introduction of parthenocarpy in Chinese cherry and Alipore rose apple by the application of growth substances. *Progressive Horticulture*, **14**(1): 61-62.

MUKHOPADHYAY, P.; GANGOPADHYAY, D.; MAZUMDAR, B.C. (2002). Further studies on qualitative constituents of bael (*Aegle marmelos* Correa) fruits grown in the *Sunderbans* area of West Bengal. *Indian Agriculturist*, **46** (1&2): 97-102.

MURUGESH, M.; VANANGAMUDI, K.; PARTHIBAN, K.T.; BHAVANISANKAR, K.; UMARANI, R.; BALAJI, B. (1998). Effect of growth regulators on germination and seedling vigour of *Emblica officinalis*. *Van Vigyan*, **36**(1): 12-14.

NACIF, S.R.; GUARDIA, M.C.; MORAES, P.I.R. (1996). [Morphology and anatomy of acerola (*Malpighia glabra* L. Malpighiaceae) seeds.] *Revista*, **48**(249): 597-610.

NAHAR, N.; RAHMAN, S.; MOSIHUZZAMAN, M. (1990). Analysis of carbohydrates in seven edible fruits of Bangladesh. *Journal of the Science of Food and Agriculture*, **51**(2): 185-192.

NAIDU, T.C.M.; SREEKANTH, B.; NARAYANAN, A. (1998). Growth and development of cashew fruit. I. Dry matter accumulation in different components of fruit. *Journal of Plantation Crops*, **26**(1): 89-92.

NAIM, Z.; KHAN, M. A.; NIZAMI, S. S. (1988). Isolation of a new isomer of ursolic acid from fruits and leaves of *Carissa carandas*. *Pakistan Journal of Scientific and Industrial Research*, **31**(11): 753-755.

NAYAK, G.; SEN, S.K. (1999). Effect of growth regulators, acid and mechanical scarification on germination of bael (*Aegle marmelos* Correa). *Environment and Ecology*, **17**(3): 768-769.

NEHRA, N.S.; GODARA, N.R.; DABAS, A.S. (1985). A note on evaluation of two variants in *phalsa* (*Grewia subinaequalis* D.C.). *Haryana Journal of Horticultural Sciences*, **14**(1/2):-58-60.

NEOG, M.; MOHAN, N.K. (1993). Physico-chemical changes during growth and development of dillenia (*Dillenia indica* Linn). *South Indian Horticulture*, **41**(2): 115-116.

NIKAWELA, J.K.; ABEYSEKARA, A.M.; JANSZ, E.R. (1998). Flabelliferins – steroidal saponins from palmyrah (*Borassus flabellifer* L.) fruit pulp. I. Isolation by flash chromatography, quantification and saponin related [foaming, haemolytic] activity. *Journal of the National Science Council of Sri Lanka*, **26**(1): 9-18.

NORMAND, F. (1994). [Strawberry Guava, relevance for Réunion], **49**(3): 217-227.

OGUNTIMEIN, B.O. (1987). The terpenoids of *Annona reticulata*. *Fitoterapia*, **58**(6): 411-413.

OGUNTIMEIN, B.O.; ELAKOVICH, S.D. (1991). Allelopathic activity of the essential oils of Nigerian medicinal plants. *International Journal of Pharmacognosy*, **29**(1): 39-44.

O'HARE, T.J. (1993). Postharvest physiology and storage of carambola (starfruit): a review. *Postharvest Biology and Technology*, **2**(4): 257-267.

OHSAWA, K.; KATO, S.; HONDA, H.; YAMAMOTO. I.
    (1990). [Pesticidal active substances in tropical plants
    – insecticidal substance from the seeds of Annonaceae.]
    Journal of Agricultural Science, Tokyo Nogyo Daigaku,
    34(4): 253-258.

OLVEIRA, M.E.B. de; OLVIERA, G.S.de; MAIA, G.A.;
    MOREIRA, R. de A.; MONTEIRO, A.C.de O. (2002).
    [Main free amino acids in cashew fruit juice: variation
    during harvest.] *Revista Brasileira de Fruitchlhera*, **24**(1):
    133-137.

OWAIYE, A.R. (1996). The yield potential of cashew
    pseudoapple colour types – a preliminary comparative
    assessment. *Discovery and Innovation*, **8**(1): 7-10.

PANDE, N.C.; SINGH, A.R.; MAURYA, V.N.; KATIYAR,
    R.N. (1986). Studies on the bio-chemical changes in bael
    (*Aegle marmelos* Correa) fruit. *Progressive Horticulture*,
    18(1/2): 29-34.

PAREEK, O.P.; SHARMA, S. (1993). Genetic resources of
    under-exploited fruits. In: *Advances in Horticulture*, Vol.
    1-*Fruit Crops, Part 1*, (ed. by K. L. Chadha and O. P.
    Pareek). Malhotra Publishing House, New Delhi-110
    064.

PARK SOHONG; LEE YONGBEOM (1997). [Effect of light
    acclimatization on photosynthetic activity of foliage
    plants. *Journal of the Korean Society for Horticultural
    Science*, **38**(1): 71-76.

PARMAR, C. (1976). Pollination and fruit set in *phalsa*
    (*Grewia asiatica* L.). *Agriculture and Agro-Industries
    Journal*, **9**(6): 12-14.

PARTHASARATHI GHOSH; PRABAL S.; SRABANI DAS;
    SWAPNADIP   THAKUR; KOKKE, W.C.M.C.;
    AKIHISA, T.; SHIMIZU, N.; TAMURA, T.;
    MATSUMOTO, T. (1991). Tyramine derivatives from
    the fruit of *Limonia acidissima. Journal of Natural Products*,
    54(5). 1389-1393.

PARUL GUPTA; VIDYA PATNI; KANT, U. (1997). *In vitro* shoot differentiation in *Emblica officinalis* Gaertn. *Journal of Phytological Research*, **7**(2): 171-172.

PATEL, M.K.; ALLAYYANAVARAMATH, S.B.; KULKARI, Y.S. (1953). Bacterial shot-hole and fruit canker of *Aegle marmelos* Correa. *Current Science*, **22**: 216-217.

PATHAK, R.K.; OJHA, C.M.; DWIVEDI, R.; HARI OM (1991). Studies on the effect of methods and duration of budding in aonla. *Indian Journal of Horticulture*, **48**(3): 207-212.

PATHAK, R.K.; PATHAK, R.A. (1993). Improvement of minor fruits. In: *Advances in Horticulture, Vol. 1-Fruit Crops, Part 1*, (ed. by K.L. Chadha and O.P. Pareek). Malhotra Publishing House, New Delhi-100 064.

PING SHENG (1999). [Xiangmi Yangtao, a high quality carambola variety.] *South China Fruits*, **28**(6): 35.

PINO, J.A. (1997). [The volatile constituents of tropical fruits. III. Feijoa and cashew.] *Alimentaria*, **35**(286): 41-45.

PINO, J.A. (1997). [The volatile constituents of tropical fruits. IV. Kiwifruit, carambola and mangosteen.] *Alimentaria*, **35**(286): 47-50.

PINO, J.A. (2000). Volatile components of Cuban *Annona* fruits. *Journal of Essential Oil Research*, **12**(5): 613-616.

PINO, J.A.; ROSADO, A.; RONCAL, E.; MARBOT, R.; AGUERO, J.; GONZALEZ, G. (1998). [Study of the components responsible for the aroma and flavour of fruits of the genus *Annona*.] *Alimentaria*, No. **298**: 81-85.

PITRI, S.; SRIVASTAVA, S.K. (1987). Pharmacological, microbiological and phytochemical studies on roots of *Aegle marmelos*. *Fitoterapla*, **58**(3): 194-197.

POI, A.K. (1989). *Studies on the Improvement of Cape-gooseberry Crop by Standardization of Some Agro-practices and Application of Different Plant Growth Regulants*. Ph.D. thesis submitted to the B.C. Krishi Vishwavidyalay, Mohanpur, W. Bengal, under the supervision of research by Dr. B.C. Mazumdar.

POI, A.K.; MAZUMDAR, B.C. (1989). Auxin-induced root regeneration in stem cuttings of Kath Bael (*Limonia acidissima*). *Indian Biologist*, **21**(1): 51-52.

POI, A.; MAZUMDAR, B.C.; SENGUPTA, P.K. (1987). Suppression of little leaf disease of cape-gooseberry by tetracycline treatment. *Indian Journal of Mycological Reserch*, **25**(2): 153-155.

PRASAD, I.D.; SENGUPTA, B.N.; SINGH, R.K.; SINGH, S.P. (1985). Effect of NPK on yield attributes and quality of cape-gooseberry (*Physalis peruviana*). *Haryana Journal of Horticultural Sciences*, **14**(3/4): 151-155.

PRATIMA ROY; MAZUMDAR, B.C. (1989). Pectin content in some minor and fruit parts. *Science and Culture*, **55**(3): 110-111.

PUROHIT, S.D.; SINGHVI, A.; TAK, K. (1996). Biochemical characteristics of differentiating callus cultures of *Feronia limonia* L. *Acta Physiologiae Plantarum*, **18**(1): 47-52.

RAO, L.J.; REDDY, M.G.R. (1989). Effect of time and severity of pruning of yield of *phalsa* (*Grewia asiatica* L.). *South Indian Horticulture*, **37**(2): 115-117.

RAO, S.S.R.; CHARYULU, N.V.N. (1989). Physiological changes during development and ripening of Syzygium cuminii (L.) Skeels. Syn. *Eugenia jambolana* Lamk. Fruits. *Indian Botanical Reporter*, **8**(1): 31-34.

RAMESH TIWARI; KALPANA DIXIT; UPADHYAY, P.S. (1987). Fungitoxicity in leaves of some higher plants against some storage fungi. *National Academy of Sciences, India, Science Letters*, **10**(12): 419-421.

RANJEET SINGH; MAHALAKSHMI, R.; VIJAAYA-CHANDRAN, S.N. (1996). The lime butterfly: potential pest of *Aegle marmelos* Corr. in nursery and its management with neem. *Insect Environment*, **2**(3): 70.

RANJIT SINGH (1969). *Fruits*. National Book Trust, India, New Delhi.

RAY, D.P.; CHATTARJEE, B.K. (1996). Effect of different concentrations of growth regulator, etiolation and vigoration treatments on the rooting of stem cutting of bael (*Aegle marmelos* Correa). *Orissa Journal of Horticulture*, **24**(1-2): 36-41.

REMA, J.; KRISHNAMOORTHY, B. (1994). Vegetative propagation of clove, *Eugenia caryophyllus* (Sprengel) B and H. *Tropical Agriculture*, **71**(2): 144-146.

ROY, G.C.; MAZUMDAR, B.C. (1989). Qualitative improvement of cashew apples and nuts by spraying with water and zinc sulphate solutions, in the coastal area of West Bengal. *Indian Journal of Landscape System and Ecological Studies*, **12**(2): 95-98.

ROY, P.K.; RAHMAN, M.; ROY, S.K. (1996). Mass propagation of *Syzygium cuminii* from selected elite trees. In: *Proceedings of the XIIIth International Symposium on Horticultural Economics, August 4-9, 1996, Rutgers, the State University of New Jersey, New Brunkswick, New Jersey, USA* [edited by Brumfield, R.G.]. *Acta Horticulturae*, No. **429**: 489-495.

ROY, S.K.; SINGH, R.N. (1978). Studies on utilization of bael fruit (*Aegle marmelos* Correa) for processing. I. Physico-chemical characteristics of different cultivars. *Indian Food Packer*, **32**: 3-8.

SAINI, J.P.; DUBE, S.D. (1992). Effect of storage on germination, tube elongation and fertilization ability of wild date pollen grains. *Progressive Horticulture*, **21**(3-4): 305-307.

SANABRIA, C.M.E.; CAMMINO, A.J.M.; GARCIA, G.; RENAUD. J. (1999). [Leaf morphology of *Monstera deliciosa* Liebm. (*Araceae*) during ontogeny.] *Ern.*, **9**(2): 103-114.

SANJAY TYAGI; MISRA, K.K.; JAISWAL, H.R. (1999). Effect of auxins on rooting of softwood stem cuttings of Carissas under mist. *Scientific Horticulture*, **6**: 37-43.

SANKAT, C.K.; BALKISSOON, F. (1992). The effect of packaging and refrigeration on the shelf life of the carambola. *ASEAN Food Journal*, **7**(2): 114-117.

SANKER, K.B.; VEERA RAGAVATHATHAM, D.; CHEZHIAN, N.; VIJAYAKUMAR, R.M.; BALASUBRAMANIAN, A. (1999). BSRI a high yield amla variety for different agroclimatic regions of Tamil Nadu. *South Indian Horticulture*, **47**(1/6): 143-144.

SANTANA, G.E.; ANGARITA, A. (1997). [Adventitious regeneration in somaclones of cape gooseberry (*Physalis peruviana*).] *Agronomla Colombiana*, **14**(1): 59-65.

SARANGI, D.; SARKAR, T.K.; ROY, A.K.; JANA, S.C.; CHATTOPADHYAY, T.K. (1992). Physico-chemical changes during growth of cape gooseberry fruit (*Physalis peruviana* L.). *Progressive Horticulture*, **21**(3-4): 225-228.

SARKAR, A.K.; DHUA, R.S.; SEN, S.K. (1984). Interaction of phenolic compounds with IBA and NAA in the regeneration of roots in water apple (*Syzygium javanicum* L.) stem cuttings. *Progressive Horticulture*, **16**(1/2): 12-15.

SARKAR, G.K.; SINGH, M.M.; MISRA, R.S. (1985). Nutritional studies in aonla (*Emblica officinalis* Gaertn.) cv. Banarasi. *Progressive Horticulture*, **17**(1): 41-46.

SCHELSTRAETE, A.; BEEL, E. (1985). [Pot culture of *Monstera deliciosa* L.] *Ver. voor. de Belgi. Ster.*, **29**(7): 313-317.

SCHEER, C.; SCHREINER, M.; HUYSKENS KEIL, S.; LUDDERS, P. (1999). [Correlation between fruit skin colour and flavour in *Physalis peruviana* L.] In *Deutsche Ges. Fur Qual. (Pfl. Nah.) DGQ e. V. XXXIV Vor. Zer. Qual., Fre., Germany, 22-23 March, 1999.*

SCHMEDA HIRSCHMANN, G.; THEODULOZ, C.; FRANCO, L.; FERRO, B.E.; ARIAS, A.R.De (1987). Preliminary pharmacological studies on *Eugenia uniflora* leaves: Xanthine oxidase as inhibitory activity. *Journal of Ethnopharmacology,* **21**(2): 183-186.

SEEMA BHADAURIA; PAHARI, G.K.; SUDHIR KUMAR (2000). Effect of *Azospirillum* biofertilizer on seedling growth and seed germination of *Emblica officinalis. Indian Journal of Plant Physiology,* **5**(2): 177-179.

SEHGAL, C.B.; KHURANA, S. (1985). Morphogenesis and plant regeneration from cultured endosperm of *Emblica officinalis* Gaertn. *Plant Cell Reports,* **4**(5): 263-266.

SEKAR, T.; FRANCIS, K. (1998). Some plant species screened for energy, hydrocarbons and phytochemicals. *Bioresource Technology,* **65**(3): 257-259.

SEKIYA, R.F.M.; CUNHA, R.J.P. (1999). [Preparation and storage effects on carambola seed germination.] *Revista Brasileira de Fruticultura,* **21**(1): 57-59.

SEN, S.K.; DE, R.K.; BANDYOPADHYAY, A. (1990). Effect of preconditioning stock plants and exogenous application of growth regulators on rooting of semi-hardwood cuttings of aonla (*Emblica officinalis* Gaertn.). *Advances in Plant Sciences,* **3**(2): 195-199.

SHAM SINGH; KRISHNAMURTHY, S.; KATYAL, S.L. (1967). *Fruit Culture in India.* Indian Council of Agricultural Research, New Delhi.

SHARMA, J.; BANDYOPADHYAY, A.; SEN, S.K. (1989). Effect of auxinic and non-auxinic chemicals on rooting or rose apple (*Syzyglum jambos* Alston) stem cuttings. *South Indian Horticulture,* **37**(2): 108-111.

SHARMA, S.S.; SHARMA, R.K.; YAMDAGNI, R. (1989). Studies on the yield and quality of fruit of aonla cultivars (*Emblica officinalis* Gaertn.) under rainfed conditions of Haryana. *Research and Development Reporter*, **6**(1): 41-43.

SHÜ ZENHONG (1999). Position on the tree affects fruit quality of bald-cut wax apples. *Journal of Applied Horticulture (Lucknow)*, **1**(1): 15-18.

SHÜ ZENHONG; CHU CHENGCHUNG; HWANG LEEJUAN; SHEIEH CHINGSHUNG. (2001). Light, temperature, and sucrose affect colour, diameter and soluble solids of disks of wax apple fruit skin. *HortScience*, **36**(2): 279-281.

SINGH, A.K.; PRABHAKAR SINGH (1998). Power of significance of difference among fruit and seed size parameters of karonda (*Carrisa carandas* Linn.). *Annals of Agricultural Research*, **19**(1) 66-71.

SINGH, G.K.; PARMAR, A.S. (1998). Studies on the effects of methods and dates of budding in aonla (*Emblica officinalis* Gaertn). *Annals of Arid Zone*, **37**(2): 199-201.

SINGH, I.S.; PATHAK, R.K. (1987). Evaluation of aonla (*Emblica officinalis* Gaertn) varieties for processing. *Acta Horticulturae*, No. **208**: 173-177.

SINGH, J.; TIWARI, R.K.S. (1998). Germination studies in karaunda (*Carissa carandas* Linn.). *Advances in Plant Sciences*, **11**(1): 313-315.

SINGH, J.N.; SINGH, S.P.; LAL BAHADUR (1989). Determination of maturity standards of aonla (*Emblica officinalis* Gaertn.) cultivars under eastern conditions of U.P. *Haryana Journal of Horticultural Science*, **18**(3-4): 216-220.

SINGH, L.B. (1954). Propagating bael (*Aegle marmelos*) vegetatively. *Science and Culture*, **19**: 405-406.

SINGH, L.B. (1961). Some promising selections of *bael*. *Annual Report of the Horticultural Research Institute*, Saharanpur.

SINGH, R.D. (1986). Studies on the bearing habit and fruit development in bael (*Aegle marmelos* Correa) cv. Mirzapuri Kagzi. *Progressive Horticulture*, **18**(3-4): 277-284.

SINGH, R.N.; ROY, S.K. (1984). The *Bael, Cultivation and Processing*. Indian Council of Agricultural Reseach, New Delhi.

SINGH, R.P.; GAUR, G.S. (1989). Growth and yield of *phalsa* (*Grewia subinaequalis* D. C.) as affected by N, P, K. *Haryana Journal of Horticultural Sciences*, **18**(1-2): 40-45.

SINGH, R.R.; JOON, M.S.; DAULTA, B.S. (1984). A note on physico-chemical characteristics of fruit in two cvs. of aonla (*Phyllanthus emblica* Linn.). *Haryana Journal of Horticultural Sciences*, **13**(3/4): 133-134.

SINHA ROY, S. (1996). *Assessment of Some Agricultural Waste Materials* as *Sources of Pectin and Changes of Pectic Compounds in Some Minor Fruits During their Development*. Ph.D. thesis submitted to the Calcutta University, under the supervision of research by Dr. B.C. Mazumdar.

SINHA ROY, S.; MAZUMDAR, B.C. (1996).Potentiality of the berries of *Carissa carandas* L. in making human consumable jelly. *Indian Biologist*, **28**(2): 19-20.

SIRCAR, P.K. (1986). A study of the rooting behavior in forced lateral cuttings. *Plant Propagator*, **32**(1): 9-11.

SITI HALIJAH ALI; MD. YUNUS JAAFAR (1992). Effect of harvest maturity on physical and chemical characteristics of carambola (*Averrhoa carambola* L.). *New Zealand Journal of Crop and Horticultural Science*. **20**(2): 133-136.

SUGIYAMA. T. (1988). [Notes on Chinese medicinal herbs and vegetables in agriculture. 5.] *Agriculture and Horticulture*, **63**(11): 1256-1258.

SULTAN SINGH; SINGHROT, R. S. (1984). Studies on the propagation of jaman (*Syzygium cumini* Skeels). I. Effect of sowing depth on seed germination and seedling growth. *Haryana Journal of Horticultural Science*, **13**(3/4): 123-126.

SULTAN SINGH; SINGHROT, R.S. (1985). A note on the effect of sowing times on jaman seed (*Syzygium cuminii* Skeel) germination and seedlings growth. *Haryana Journal of Horticultural Sciences*, **14**(3/4): 215-217.

SUPE, V.S.; SHETE, M.B.; CHAVAN, U.D.; KAULGUD, S.N. (1998 b). Physico-chemical analysis of different aonla (*Emblica officinalis*) cultivars under Maharashtra conditions. *Journal of Maharashtra Agricultural Universities*, **22**(3): 310-312.

SUSUMU ARIMA; DAIGOHO, M.; HOQUE, W.A. (1999). Flower development and anthesis behviour in the water chestnut (*Trapa* sp.). *Bulletin of the Faculty of Agriculture Saga University*, No. **84**: 83-92.

TRINCHERO, G.D.; SOZZI, G.O.; CERRI, A.M.; VILELLA, F.; FRASCHINA, A.A. (1999). Ripening-related changes in ethylene production, respiration rate and cell-wall enzyme activity in goldenberry (*Physalis peruviana* L.), a solanaceous species. *Postharvest Biology and Technology*, **16**(2): 139-145.

UTHAIAH, B.C. (1988). Studies on changes in physical parameters of the developing wild edible fruit *Carissa carandas*. *Myforest*, **24**(4): 256-258.

VARGHESE, S.K.; INAMDAR, J.A.; KIRAN KALIA; SUBRAMANIAN, R.B.; NATARAJ, M. (1993). Micropropagation of *Aegle marmelos* (L.) Corr. *Phytomorphology*, **43**(1/2): 87-92.

VEERASAMY, S.; RAJARATHINAM, K.; JAYABALAN, M. (1995). Regeneration of shoots from split shoot apical meristem in *Borassus Flabellifer*. *Journal of Ecotoxicology and Environmental Monitoring*, **5**(1): 39-43.

VELU, G. (1989). Impact of organic manuring of palmyrah. *Madras Agricultural Journal*, **76**(10): 592-598.

VENDRAMINI, A.L.; TRUGO, L.C. (2000). Chemical composition of acerola fruit (*Malpighia punicifolia* L.) at three stages of maturity. *Food Chemistry*, **71**(2): 195-198.

VENKATARATNAM, L. (1965). *Sitaphal and Other Annona Fruits in India*. Farm Information Unit, Directorate of Extension, Ministry of Food and Agriculture, New Delhi.

VENKITAKRISHNAN, M.; PANJATCHARAM, V.; KUMARAVELU, G.; RAMANUJAM, M.P. (1997). Physico-chemical changes during maturation and ripening of jambolan fruit. *Indian Journal of Plant Physiology* **2**(4): 267-270.

VERNIN, G.; VERNIN, C.; PIERIBATTESTI, J.C.; ROQUE, C. (1998). Analysis for the volatile compounds of *Psidium cattleianum* Sabine fruit from Réunion island. *Journal of Essential Oil Research*, **10**(4): 353-362.

VISENTAINER, J.V.; VIEIRA, O.A.; MATSUSHITA, M.; SOUZE, N.E. DE (1997). Physico-chemical characterization of acerola (*Malpighia glabra* L.) produced in Maringá. Parana State, Brazil. *Archivos Latinoamericanos de Nutrición*, **47**(1): 70-72.

VOLTOLINI, J.A.; FACHINELLO, J.C. (1997). Effect of shading Cattley guava stock plant (*Psidium cattleyanum* Sabina) on propagation by cuttings. *Acta Horticulturae*, No. **452**: 59-62.

WAGH, A.P.; CHOUDHARY, M.H.; KULWAL, L.V.; JADHAV, B.J.; JOSHI, P.S. (1998). Effect of seed treatment on germination of seed and initial growth of aonla seedling in polybag. *PKV Research Journal,* **22**(2): 176-177.

WANG JIAFU; HE BIZHU (2000). [*In vitro* culture of longan shoot tips.] *Journal of Fujian Agricultural University,* **29**(1): 23-26.

WATTANAWIKKIT, P.; TANTIWIWAT, S.; SURAWATANANON, S.; SANGTHONGPROW, S. (1999). [*In vitro* culture of Malay apple (*Engenia malaccensis* Linn.).] In: *The 37th Kasetsart University Annual Conference, 3-5 February, 1999 (edited by Outes, C. G.)* Bangkok, Thailand. Text and Journal Publication Co. Ltd., 235-240.

WOLFF, X.Y. (1991). Species, cultivar, and soil amendments influence fruit production of two *Physalis* species. *HortScience,* **26**(12): 1558-1559.

WONG, K.C.; KHOO, K.H. (1993). Volatile components of Malaysian *Annona* fruits. *Flavour and Fragrance Journal,* **8**(1): 5-10.

WONG, K.C.; LAI, F.Y. (1996). Volatile constituents from the fruits of four *Syzygium* species grown in Malaysia. *Flavour and Fragrance Journal,* **11**(1): 61-66.

WONG, K.C.; WONG, S.N.; LOI, H.K.; LIM, C.L. (1996). Volatile constituents from the fruits of four edible Sapindaceae: rambutan (*Nephelium lappaceum* L.), pulasan [*N. ramboutan-ake* (Labill.)], longan (*Dimocarpus longan* Lour.), and mata kucing (*D. longan* spp. *malesianus* Leenh.). *Flavour and Fragrance Journal,* **11**(4): 223-229.

WU DING YAO; QIU JINDAN; ZHANG HAILAN; LUO XIAOZHENG (2000). [A study on flowering promotion by ringing in longan (*Dimocarpus longana* Lour.)] *Scientia Agricultura Sinica,* **33**(6) 40-43.

XIA, Q.H.; CHEN, R.Z.; FU, J.R. (1992 a). Effects of desiccation, temperature and other factors on the germination of lychee (*Litchi chinensis* Sonn.) and longan (*Euphoria longan* Steud.) seeds. *Seed Science and Technology*, **20**(1): 119-127.

XIA, Q.H.; CHEN, R.Z.; FU, J.R. (1992 b). Moist storage of lychee (*Litchi chinensis* Sonn.) and longan (*Euphoria longan* Steud.) seeds. *Seed Science and Technology*, **20**(2): 269-279.

XU XIUDAN; HUANG JINSONG; ZHENG SHAOQUAN; LU XIUMIN; XU JIAHUI (1999). [An extremely late maturing longan variety 'Lidongben'.] *Acta Horticulturae Sinica*, **26**(2): 136.

YADAV, H.S. (1992). Efficacy of monocrotophos against *Singhara* blood worm *Stenochironomus* spp. *Indian Journal of Plant Protection*, **20** 1): 58-60.

YADAV, H.S.; GARGAV, V.P. (1988). Chemical control of singhara beetle, *Galerucella birmanica* Jacoby. *Indian Journal of Plant Protection*, **16**(2): 159-162.

YADAV, U.; LAL, M.; JAISWAL, V.S. (1990). *In vitro* micropropagation of the tropical fruit tree *Syzygium cuminii* L. *Plant Cell, Tissue and Organ Culture*, **21**(1): 87-92.

YADAV, V.K.; SINGH, H.K. (1999). Studies on the preharvest application of chemicals on shelf-life of aonla (*Emblica officinalis* Gaertn.) fruits at ambient temperature. *Journal of Applied Horticulture* (*Lucknow*), **1**(2): 118-121.

YAMADA, M.; HIDAKA, T.; FUKAMACHI, H. (1996). Heat tolerance in leaves of tropical fruit crops as measured by chlorophyll fluorescence. *Scientia Horticulturae*, **67**(1/2): 39-48.

YANG WINHUAN (1996). [Songfengben, an uncommon and promising late longan variety.] *China Fruits*, No. **1**: 52.

YEH DERMING (1998). [Growth of six foliage plants in subirrigation systems.] *Journal of the Chinese Society for Horticulture Science*, **44**(1): 81-92.

ZANG XIAOPING; XU NENGKUDN (2000). [On the fertilizing for custard apple trees.] *South China Fruits*, **29**(3): 29-30.

ZARGA, M.H.A. (1986). Three new simple indole alkaloids from *Limonia acidissima. Journal of Natural Products*, **49**(5): 901-904.

ZEIER, J.; SCHREIBER, L. (1998). Comparative investigation of primary and tertiary endodermal cell walls isolated from the roots of five monocotyledoneous species: chemical composition in relation to fine structure. *Planta*, **206**(3): 349-361.

ZHOU, J.W.; ZHANG, Z.P.; ZHOU, G.Y.; LIU, S.Z.; CHIN, B.Q. (1983). [Culture of axillary buds of Nanhu Singhara nut.] *Plant, Physiology Communications, (Zhiwu Shenglinue Tongxun)* No. **1**: 31.

ZHUANG YIMEI; WANG RENJI; XIE ZHINAN; XI WENBAO (1995). [Optimum range of mineral element contents in the leaves of Shuizhang longan.] *Journal of Fujian Agricultural University*, **24**(3): 281-286.

ZORA SINGH; GREWAL, G. P. S.; LAKHVIR SINGH (2000). Effects of gibberellin A_4/A_7 and blossom thinning on fruit set, retention, quality, shoot growth and return bloom of phalsa (*Grewia asiatica* L.). *Acta Horticulturae*, No. **525**: 463-466.